MODELLING SUBCOOLED BOILING FLOWS

MODELLING SUBCOOLED BOILING FLOWS

G. H. YEOH AND J. Y. TU

Nova Science Publishers, Inc.
New York

Copyright © 2009 by Nova Science Publishers, Inc.

All rights reserved. No part of this book may be reproduced, stored in a retrieval system or transmitted in any form or by any means: electronic, electrostatic, magnetic, tape, mechanical photocopying, recording or otherwise without the written permission of the Publisher.

For permission to use material from this book please contact us:
Telephone 631-231-7269; Fax 631-231-8175
Web Site: http://www.novapublishers.com

NOTICE TO THE READER
The Publisher has taken reasonable care in the preparation of this book, but makes no expressed or implied warranty of any kind and assumes no responsibility for any errors or omissions. No liability is assumed for incidental or consequential damages in connection with or arising out of information contained in this book. The Publisher shall not be liable for any special, consequential, or exemplary damages resulting, in whole or in part, from the readers' use of, or reliance upon, this material.

Independent verification should be sought for any data, advice or recommendations contained in this book. In addition, no responsibility is assumed by the publisher for any injury and/or damage to persons or property arising from any methods, products, instructions, ideas or otherwise contained in this publication.

This publication is designed to provide accurate and authoritative information with regard to the subject matter covered herein. It is sold with the clear understanding that the Publisher is not engaged in rendering legal or any other professional services. If legal or any other expert assistance is required, the services of a competent person should be sought. FROM A DECLARATION OF PARTICIPANTS JOINTLY ADOPTED BY A COMMITTEE OF THE AMERICAN BAR ASSOCIATION AND A COMMITTEE OF PUBLISHERS.

LIBRARY OF CONGRESS CATALOGING-IN-PUBLICATION DATA
   Yeoh, Guan Heng.
    Modelling subcooled boiling flows / G. H. Yeoh and J. Y. Tu.
      p. cm.
    ISBN 978-1-60456-943-8 (softcover)
    1. Heat--Transmission--Mathematical models. 2. Fluid dynamics. I. Tu, Jiyuan. II. Title.
    TJ260.Y46 2008
    621.402'2015118--dc22
                                              2008033027

Published by Nova Science Publishers, Inc. ✦ *New York*

CONTENTS

Preface		vii
Chapter 1	Introduction	1
Chapter 2	Conservation Equations for the Two-Fluid Boiling Model	5
Chapter 3	Numerical Methods for Subcooled Boiling Flow	11
Chapter 4	Application of Empirical Relationships to Modeling Subcooled Boiling Flow	19
Chapter 5	Phenomenological Observations and Population Balance Model	37
Chapter 6	Improvements to Wall Heat Flux Partitioning Model	49
Chapter 7	Summary	71
References		73
Index		81

PREFACE

In the context of computational fluid dynamics (CFD), modeling low-pressure subcooled boiling flow is of particular significance. A review is provided in this article of the various numerical modeling approaches that have been adopted to handle subcooled boiling flow. The main focus in the analysis of such a challenging problem can be broadly classified according into two important categories: (i) Heat transfer and wall heat flux partitioning during subcooled boiling flow at the heated wall and (ii) Two-phase flow and bubble behaviors in the bulk subcooled flow away from the heated wall. For the first category, details of both empirical and mechanistic models that have been proposed in the literature are given. The enhancement in heat transfer during forced convective boiling attributed by the presence of both sliding and stationary bubbles, force balance model for bubble departure and bubble lift-off as well as the evaluation of bubble frequency based on fundamental theory depict the many improvements that have been introduced to the current mechanistic model of heat transfer and wall heat flux partitioning. For the second category, details of applications of various empirical relationships and mechanistic model such as population balance model to determine the local bubble diameter in the bulk subcooled liquid that have been employed in the literature are also given. A comparison of the predictions with experimental data is demonstrated. For the local case, the model considering population balance and improved wall heat partition shows good greement with the experimentally measured radial distributions of the Sauter mean bubble diameter, void fraction, interfacial area concentration and liquid velocity profiles. Significant weakness prevails however over the vapor velocity distribution. For the axial case, good agreement is also achieved for the axial distributions of the Sauter mean bubble diameter, void fraction and interfacial area concentration profiles. The present model correctly represents the plateau at the initial boiling stages at upstream, typically found in low-pressure subcooled boiling flows, followed by the significant rise of the void fraction at downstream.

NOMENCLATURE

a_{if}	interfacial area concentration
A_B	bubble area
A_C	cross-sectional area of boiling channel
A_q	fraction of heater area occupied by bubbles
ΔA	discrete area of control volume
Bo	boiling number = Q_w/Gh_{fg}
c_f	constant in equation (44)
C_1, C_2	constants defined in equation (84)
C_{bw}	constant in equation (57)
C_D	drag coefficient
C_L	shear lift coefficient
C_{TD}	turbulent dispersion coefficient
C_p	specific heat
C_s	constant defined in equation (80)
C_{w1}, C_{w2}	wall lubrication constants
C_μ	turbulent constant = 0.09
$C_{\mu b}$	bubble induced turbulent constant
C_v	acceleration coefficient
d	vapor bubble diameter at heated surface
d_{bw}	bubble departure diameter
$d_{b,max}$	maximum allowed bubble diameter
d_i, d_j	daughter particle diameters
d_{ij}	equivalent diameter
d_w	surface/bubble contact diameter
D	average bubble diameter
D_b	departing bubble diameter
D_d	bubble departure diameter
D_l	bubble lift-off diameter
D_s	Sauter bubble diameter
D_{B1}, D_{B2}	reference bubble diameters
D_B	death rate due to break-up
D_C	death rate due to coalescence

f	particle size distribution or bubble departure frequency
f_i	scalar fraction related to the number density of the discrete bubble classes
F	degree of surface cavity flooding
F_b	buoyancy force
F_h	force due to the hydrodynamic pressure
F_{du}	unsteady drag force due to asymmetrical growth of the bubble
F_s	surface tension force
F_{sL}	shear lift force
F_x	forces along the x-direction
F_y	forces along the y-direction
F_{lg}	action of interfacial forces from vapor on liquid
F_{gl}	action of interfacial forces from liquid on vapor
F_{lg}^{drag}	drag force
F_{lg}^{lift}	lift force
$F_{lg}^{lubrication}$	wall lubrication force
$F_{lg}^{dispersion}$	turbulent dispersion force
g	gravitational constant
$\vec{g}$	gravitational vector
G	mass flux
G_s	dimensionless shear rate
h	interfacial heat transfer coefficient
h_0	initial film thickness
h_f	critical film thickness at rupture
h_{fg}	latent heat of vaporization
H	enthalpy
Ja	Jakob number based on wall superheat
Ja_{sub}	Jakob number based on liquid subcooling = $\rho C_{pl}(T_{sat}-T_l)/\rho_g h_{fg}$
k	turbulent kinetic energy
K	projected area of bubble
l_s	sliding distance
$\vec{n}$	normal to the surface
n	number density
n_i	number density of the discrete bubble ith class
n_j	number density of the discrete bubble jth class

N	number of bubble classes
N_a	active nucleation site density
N_f	total number of discrete areas of a control volume element
Pe	Peclet number
P_B	production rate due to break-up
P_C	production rate due to coalescence
Q_w	wall heat flux
Q_c	heat transfer due to forced convection
Q_e	heat transfer due to evaporation
Q_{tc}	heat transfer (transient conduction) due to stationary bubble
Q_{tcsl}	heat transfer (transient conduction) due to sliding bubble
$\vec{r}$	spatial position vector
r	bubble radius at heated wall, flow spacing within annular channel
r_c	cavity radius at heated surface
r_r	curvature radius of the bubble at heated surface
Re	flow Reynolds number = Gd_h/μ_l
Re_b	bubble Reynolds number
R_f	ratio of the actual number of bubbles lifting off to the number of active nucleation sites
R_i	radius of inner heated wall
R_o	radius of outer unheated wall
R_{ph}	source/sink term due to phase change
s	spacing between nucleation sites
S_{ϕ_k}	source term for generic variable ϕ_k
S_i	additional source terms due to coalescence and breakage
St	Stanton number
t	thermo-fluid time scale
t_g	bubble growth period
t_{ij}	coalescence time
t_{sl}	bubble sliding period
t_w	bubble waiting period
T	temperature
T_{sub}	local subcooling temperature = $T_{sat} - T_l$
$T_{sub1}\ T_{sub2}$	reference subcooling temperatures
ΔT	difference in temperature
P	pressure
u	velocity

$\bar{u}$	velocity vector in Cartesian frame = u, v, w
u_t	velocity due to turbulent collision
u_τ	friction velocity
v_i	specific volume of discrete bubble ith class
ΔV	control volume
V_b	bubble velocity
V_B	bubble volume
x	Cartesian coordinate along x
x^+	non-dimensional normal distance from heated wall
x_i	pivot size or abicissa
y	Cartesian coordinate along y
y_w	adjacent point normal to the wall surface

GREEK LETTERS

α	advancing angle
α_g	vapor void fraction
α_l	liquid void fraction
β	receding angle or measured constant in equation (58)
χ	coalescence rate
δ_l	thermal boundary layer thickness
ε	turbulent dissipation rate
ϕ_k	geneic variable
ϕ_{WN}	bubble nucleation rate
ϕ_{COND}	bubble condensation rate
η	thermal diffusivity
λ	thermal conductivity or size of an eddy
μ	viscosity
μ^{eff}	effective viscosity
θ	bubble contact angle
θ_i	inclination angle
θ_{ij}	turbulent collision rate
ρ	density
σ	surface tension
τ_{ij}	bubble contact time
ξ	size ratio between an eddy and a particle in the inertial sub-range

ξ_H	heated perimeter of boiling channel
Γ_{ϕ_k}	diffusion coefficient for generic variable ϕ_k
Γ_{lg}	interfacial mass transfer from vapor to liquid
Γ_{gl}	interfacial mass transfer from liquid to vapor
Ω	break-up rate

SUBSCRIPTS

axial	axial distribution
eff	effective
g	vapor
inlet	channel entrance
l	liquid
local	local distribution
nb	neighboring centriods
non	non-orthogonal
ONB	onset of nucleate boiling
P	centriod of control volume element
s	surface heater
sat	saturation
sub	subcooling
Tb	bubble induced turbulence
Tl	liquid shear-induced turbulence
w	wall

Chapter 1

INTRODUCTION

During the last five decades, numerous numerical models have been developed to predict the fluid flow and heat transfer characteristics of subcooled boiling flow. is the boiling process by nature is inherently complex, which still makes the analysis extremely challenging. Subcooled boiling flow can usually be characterized by the presence of thermodynamic non-equilibrium between the liquid and vapor phases. A *high-temperature* two-phase region exists near the heated wall whilst a *low-temperature* single-phase liquid prevails away from the heated surface. Consider the schematic drawing of a subcooled boiling flow accompanied by a typical boiling curve describing the void fraction distribution in Figure 1. Heterogeneous bubble nucleation occurs within the small pits and cavities designated as nucleation sites on the heated surface. These nucleation sites are activated when the temperature of the surface exceeds the saturation liquid temperature at the local pressure. Away from the wall, the temperature of the bulk fluid remains below saturation and is, by definition, subcooled. At a point called the onset of nucleate boiling (ONB), boiling occurs and bubbles remain attached to the heater surface. Further downstream, as the bulk temperature liquid temperature increases, the bubbles grow larger and begin to detach from the heater surface. The void fraction increases sharply at a location called the net vapor generation (NVG), which indicates the transitional point between two regions: low void fraction region followed by another region in which the void fraction increases significantly thereafter. This review will mainly focus on subcooled boiling flow especially at low-pressure applications.

Modeling low-pressure subcooled boiling flow in the context of computational fluid dynamics (CFD) can be broadly classified according to two categories: (i) Heat transfer and wall heat flux partitioning during subcooled boiling flow at the heated wall and (ii) Two-phase flow and bubble behaviors in the bulk subcooled flow away from the heated wall.

The first category generally entails the development of models to predict the heat transfer rate during subcooled nucleate boiling flow by appropriately partitioning the wall heat flux. The various approaches that have been adopted are empirical correlations for wall heat flux, empirical correlations for the partitioning of the wall heat flux and mechanistic models for wall heat flux and partitioning.

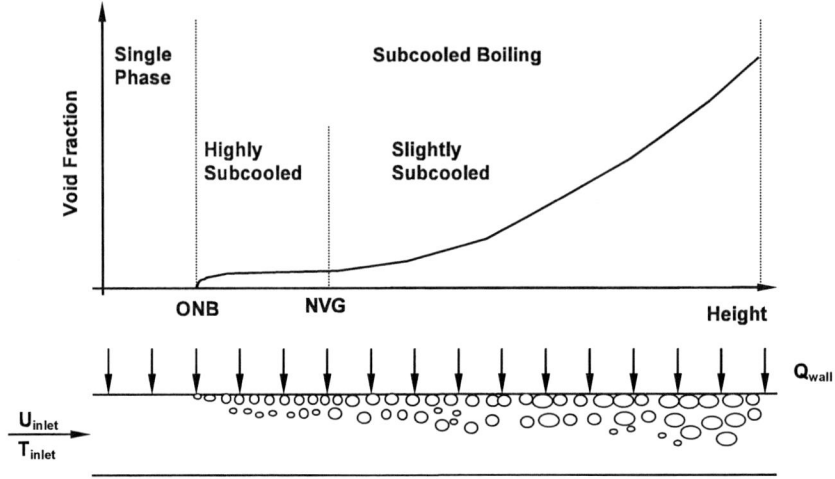

Figure 1. A schematic illustration of a subcooled boiling flow in a heated channel.

On the basis where empirical correlations for wall heat flux are employed, these correlations are usually limited to the prediction of total wall heat flux for a particular boiling flow situation. These empirical heat transfer models attempt to predict the wall heat flux given the wall superheat ΔT_w for a particular flow condition in the regions covering the singles-phase forced convection, partial nucleate boiling and full-developed boiling. In the partial nucleate boiling region, which is the transition region between single-phase forced-convection and fully-developed nucleate boiling, the correlations of Bergles and Rohsenow [1], Mikic and Roshsenow [2], Chen [3] and Kandlikar [4] are widely used. Several correlations have also been proposed for the fully-developed nucleate boiling of which specific relationships developed by McAdams et al. [5], Thom et al. [6] and Kandlikar [4] are popular expressions. Despite their extensive usage, they are still nevertheless merely correlations of experimental data and as such they preclude the modeling of heat transfer mechanisms and are unable to provide any information regarding the partitioning of the wall heat flux between the liquid and vapor phases.

A number of correlations for wall heat flux partitioning have also been proposed in the literature. These correlations differ from above, namely, what heat transfer mechanisms are considered, and what fraction of the total heat flux is transferred by each of these mechanisms. Various processes occurring at the heater surface such as the information required for the vapor generation and condensation rates are distinguished. The primary goal of the correlations is to predict the bulk void fraction; empirical correlations are formulated to provide information on how the wall heat flux is to be partitioned but not of the total wall heat flux itself. Useful models such as proposed by Griffith et al. [7], Zuber and Findlay [8], Lahey [9], Chatoorgoon et al. [10] and Zeitoun [11] are noted for their significant contributions to the study of subcooled boiling flow.

Mechanistic models based on relevant heat transfer mechanisms occurring during the boiling process can however be readily employed for both the prediction of the wall heat flux as well as the partitioning of the wall heat flux between the liquid and vapor phases. Most CFD investigations on subcooled boiling flow are mainly based on the use of these mechanistic models. In essence, the wall heat flux can be partitioned into three heat flux components: single-phase, transient conduction and evaporation. Single-phase heat transfer is assumed to occur in areas of the heated surface unaffected by the bubbles. Transient conduction (quenching heat flux component) occurs over the area of the heater surface under the influence of bubbles while the single-phase heat flux component acts over the area unaffected by bubbles. The model consists of three unknown parameters, which should be modeled accurately. They are: active nucleation site density (N_a), departing bubble diameter (D_b) and bubble frequency (f).

The second category involves the handling of the fluid flow in the bulk subcooled liquid. The two-fluid model based on the Eulerian framework represents the most detailed and accurate macroscopic formulation of the thermal and hydrodynamic characteristics of any two-phase systems. In the two-fluid model, the field equations are expressed by two sets of six conservation equations of mass, momentum and energy. For subcooled boiling flow, one set is dedicated to resolve the gas bubble phase while the other is solved for the liquid water phase. Through an appropriate averaging of the local instantaneous balance equations, phasic interaction terms appear in each of the averaged balance equations. These interfacial terms represent the mass, momentum and energy transfers through the interface between the phases. The existence of these interfacial transfer terms is rather significant as they determine the rate of phase changes, and the degree of mechanical and thermal non-equilibrium between phases. Most importantly, they provide the necessary closure relations to the two-fluid model formulation. In the mass conservation equations, mass transfer is accounted between phases due to gas bubbles being condensed in the bulk subcooled liquid. In the momentum conservation equations, the important interfacial effects between the gas and liquid phases due to the drag force as well as other possible so-called non-drag forces in the form of lift, wall lubrication and turbulent dispersion are incorporated. In the energy conservation equations, the interfacial heat transfer accounts for the phase change due to condensation.

Also, the prediction of the local bubble sizes in the subcooled liquid flow is expected to be strongly influenced by the complex bubble behaviors in the two-phase flows occurring near the heated wall. A detailed mathematical analysis indicates that empirical correlations can at best only provide a macroscopic description of the boiling phenomenon. In reality, complex mechanistic behaviors of bubble coalescence and bubble collapse due to condensation (microscopic in nature) exist in the two-phase subcooled boiling flow. Physical insights suggest that the numerical solutions could be significantly improved by accounting these bubble mechanistic behaviors especially in determining the local bubble size such as the Sauter bubble diameter. In the two-fluid model, the Sauter bubble diameter appears in the interfacial transfer terms of the mass, momentum and energy. An accurate determination of the Sauter bubble diameter is crucial as the bubble size influences the inter-phase heat and mass transfer through the interfacial area concentrations and momentum drag terms. Because of the many successes in applying the population balance approach towards better describing and understanding complex industrial flow systems [12–16], the potential to implement and extend the modeling of subcooled boiling flow to predict the non-uniform bubble size distribution is of enormous significance. It will be demonstrated later that the specific development of the population balance approach for subcooled boiling flows has contributed to a considerable improvement of the two-fluid boiling model formulation.

A review of the various models/correlations developed for the prediction of heat transfer and wall heat flux partitioning as well as the formulation of conservations equations for the two-fluid boiling model and population balance model is given in the following sections.

Chapter 2

CONSERVATION EQUATIONS FOR THE TWO-FLUID BOILING MODEL

The two-fluid model treating both the vapor and liquid phases as continua solves two sets of conservation equations governing mass, momentum and energy which can be expressed as:

Continuity Equation of Liquid Phase

$$\frac{\partial \rho_l \alpha_l}{\partial t} + \nabla \cdot \left(\rho_l \alpha_l \vec{u}_l \right) = \Gamma_{lg} \tag{1}$$

Continuity Equation of Vapor Phase

$$\frac{\partial \rho_g \alpha_g}{\partial t} + \nabla \cdot \left(\rho_g \alpha_g \vec{u}_g \right) = \Gamma_{lg} \tag{2}$$

Momentum Equation of Liquid Phase

$$\frac{\partial \rho_l \alpha_l \vec{u}_l}{\partial t} + \nabla \cdot \left(\rho_l \alpha_l \vec{u}_l \vec{u}_l \right) = -\alpha_l \nabla P + \alpha_l \rho_l \vec{g} + \nabla \cdot \left[\alpha_l \mu_l^{eff} \left(\nabla \vec{u}_l + \left(\nabla \vec{u}_l \right)^T \right) \right] \\ + \left(\Gamma_{lg} \vec{u}_g - \Gamma_{gl} \vec{u}_l \right) + F_{lg} \tag{3}$$

Momentum Equation of Vapor Phase

$$\frac{\partial \rho_g \alpha_g \bar{u}_g}{\partial t} + \nabla \cdot \left(\rho_g \alpha_g \bar{u}_g \bar{u}_g\right) = -\alpha_g \nabla P + \alpha_g \rho_g \bar{g} + \nabla \cdot \left[\alpha_g \mu_g^{eff}\left(\nabla \bar{u}_g + \left(\nabla \bar{u}_g\right)^T\right)\right]$$
$$+ \left(\Gamma_{gl}\bar{u}_l - \Gamma_{lg}\bar{u}_g\right) + F_{gl} \quad (4)$$

Energy Equation of Liquid Phase

$$\frac{\partial \rho_l \alpha_l H_l}{\partial t} + \nabla \cdot \left(\rho_l \alpha_l \bar{u}_l H_l\right) = \nabla \cdot \left[\alpha_l \lambda_l \nabla T_l + \frac{\mu_{Tl}}{Pr_{Tl}}\nabla H_l\right] + + \left(\Gamma_{lg}H_g - \Gamma_{gl}H_l\right) \quad (5)$$

Energy Equation of Vapor Phase

$$\frac{\partial \rho_g \alpha_g H_g}{\partial t} + \nabla \cdot \left(\rho_g \alpha_g \bar{u}_g H_g\right) = \nabla \cdot \left[\alpha_g \lambda_g \nabla T_g + + \frac{\mu_{Tg}}{Pr_{Tg}}\nabla H_g\right] + \left(\Gamma_{gl}H_l - \Gamma_{lg}H_g\right) \quad (6)$$

In subcooled boiling flow, the source term Γ_{lg} in equation (1) represents the mass transfer rate due to condensation in the bulk subcooled liquid. It can be expressed by:

$$\Gamma_{lg} = \frac{h a_{if}\left(T_{sat} - T_l\right)}{h_{fg}} \quad (7)$$

where h is the inter-phase heat transfer coefficient determined from Ranz and Marshall [17] Nusselt number correlation and a_{if} is the interfacial area between phases per unit volume. For mass conservation, it is noted that the source term of the mass transfer rate Γ_{gl} in equation (2) is the negative of Γ_{lg} in equation (1), *viz.*, $\Gamma_{gl} = -\Gamma_{lg}$. The wall vapor generation rate Γ_{wg} is modeled in a mechanistic way derived by considering the total mass of bubbles detaching from the heated surface as:

$$\Gamma_{wg} = \frac{Q_e}{h_{fg} + C_{pl}T_{sub}} \quad (8)$$

where Q_e is the heat transfer due to evaporation. This wall nucleation rate is usually accounted as a specified boundary condition for the continuity equation of vapor phase.

Interfacial transfer terms in the momentum and energy equations represented by Γ_{lg} and F_{lg} denote the transfer terms from the gas phase to the liquid phase. The mass transfer Γ_{lg} for subcooled boiling flow is already given in equation (7) whilst the total interfacial force F_{lg} considered in the present study includes the following forces:

$$F_{lg} = F_{lg}^{drag} + F_{lg}^{lift} + F_{lg}^{lubrication} + F_{lg}^{dispersion} \qquad (9)$$

The terms on the right-hand side in equation (9) represent the drag force, lift force, wall lubrication force and turbulent dispersion force respectively. This force strongly governs the distribution of the liquid and gas phases within the flow volume. The interfacial momentum transfer between gas and liquid due to drag force F_{lg}^{drag} is a result of the shear and form drag of the fluid flow. It is modeled according to Ishii and Zuber [18] as:

$$F_{lg}^{drag} = -F_{gl}^{drag} = \frac{1}{8} C_D a_{if} \rho_l |\vec{u}_g - \vec{u}_l|(\vec{u}_g - \vec{u}_l) \qquad (10)$$

Owing to the horizontal velocity gradient, bubbles rising in a liquid are subjected to a lateral lift force. Following Drew and Lahey [19], the lift force in terms of the slip velocity and the curl of the liquid phase velocity is described by:

$$F_{lg}^{lift} = -F_{gl}^{lift} = \alpha_g \rho_l C_L (\vec{u}_g - \vec{u}_l) \times (\nabla \times \vec{u}_l) \qquad (11)$$

In addition to the lift force, a lateral force is also formed to prevent bubbles attaching on the solid walls due to surface tension. The wall lubrication force, which is in the normal direction away from the heated wall and decays with distance from the wall, can be modeled as [20]:

$$F_{lg}^{lubrication} = -F_{gl}^{lubrication} = -\frac{\alpha_g \rho_l (\vec{u}_g - \vec{u}_l)}{D_s} \max\left(0, C_{w1} + C_{w2} \frac{D_s}{y_w}\right) \vec{n} \qquad (12)$$

To avoid the emergence of attraction force, the force is set to zero for large y_w.

For turbulence assisted bubble dispersion, turbulence induced dispersion taken as a function of turbulent kinetic energy and gradient of the liquid void fraction can be expressed in the form of [21]:

$$F_{lg}^{dispersion} = -F_{gl}^{dispersion} = -C_{TD}\rho_l \kappa \nabla \alpha_l \tag{13}$$

More detail descriptions of these forces can be found in Anglart and Nylund [22] and Lahey and Drew [23]. The drag coefficient C_D in equation (10) has been correlated for several distinct Reynolds number regions for individual bubbles according to Ishii and Zuber [18]. For the lift coefficient C_L, Lopez de Bertodano [20] and Takagi and Matsumoto [24] suggested a value of $C_L = 0.1$. Drew and Lahey [19] proposed $C_L = 0.5$ based on objectivity arguments. Tomiyama [25] developed however an Eötvos number dependent correlation that allows negative coefficients to emerge if the bubble diameter is larger than 5.5 mm (i.e. for airwater system). This results in a negative lift force forcing the large bubbles to be emigrated towards the tube centre. For the range of flow conditions considered in this paper, the bubbly flow regime persisted in the subcooled boiling flow; only positive lift forces appeared for small bubbles. The constant of $C_L = 0.01$ as suggested by Wang et al. [26] has been found to be appropriate for the wide range of boiling flows investigated. The wall lubrication constants $C_{w1} = -0.01$ and $C_{w2} = 0.05$ as suggested by Antal et al. [21] are employed while the recommended value for $C_{TD} = 0.1$ according to Kurul and Podowski [27] is adopted for the turbulent dispersion force.

Unlike single phase fluid flow problem, no standard turbulence model has been tailored for two-phase turbulent flow. In majority of two-phase flow applications, the standard two-equation k-ε turbulence model is employed to resolve the turbulent flow associated with the continuous liquid and dispersed vapor phases. For simplicity, the standard k-ε model has been employed throughout the various subcooled boiling investigations herein. It is nevertheless noted that Menter's k-ω based shear stress transport (SST) [28] could also be alternatively employed to resolve the two-phase turbulent subcooled boiling flow. In the recent work by Cheung et al. [29] and numerical calculations performed by Frank et al. [30], the "wall" peaking behavior of the void fraction distribution in an isothermal bubbly flow was seen to be better captured through the SST turbulent model. This could be attributed to a more realistic prediction of the turbulent dissipation close to the wall through the k-ω formulation. Further development should be focused on two-phase flow turbulence modeling in order to better understand or improve existing models.

On the basis of the eddy viscosity hypothesis, the shear-induced viscosity in the momentum liquid phase and vapor phase equations can be taken as the sum of the molecular viscosity and turbulent viscosity of each phase. By the inclusion of Sato's [31] model for bubble-induced turbulence, the effective viscosity is thus considered as the total of the shear-induced turbulent viscosity and the bubble-induced turbulent viscosity. The viscosity of the liquid phase can be expressed as:

$$\mu_l^{eff} = \mu_l + \mu_{Tl} + \mu_{Tb} \tag{14}$$

where the liquid turbulent viscosity is given by

$$\mu_{Tl} = \rho_l C_\mu \frac{k_l^2}{\varepsilon_l} \tag{15}$$

and the extra bubble induced turbulent viscosity is evaluated according to

$$\mu_{Tb} = \rho_l C_{\mu b} \alpha_g D_s \left| \bar{u}_g - \bar{u}_l \right| \tag{16}$$

The constant $C_{\mu b}$ has a value of 0.6. Effective viscosity in the vapor phase can now be simply evaluated as

$$\mu_g^{eff} = \mu_l^{eff} \frac{\rho_g}{\rho_l} \tag{17}$$

Chapter 3

NUMERICAL METHODS FOR SUBCOOLED BOILING FLOW

BASIC ASPECTS OF DISCRETISATION – FINITE VOLUME METHOD

It can be observed that there are commonalities between the various conservation equations for the two-fluid boiling model. Employing the general variable ϕ_k, the generic form of the governing equations in Chapter 2 can be written in the form as:

$$\frac{\partial(\rho_k \alpha_k \phi_k)}{\partial t} + \nabla \cdot (\rho_k \alpha_k \vec{u}_k \phi_k) = \nabla \cdot \left[\alpha_k \Gamma_{\phi_k}^k \nabla \phi_k \right] + S_{\phi_k}^k \tag{18}$$

Equation (18) is aptly known as the transport equation for any variables ϕ_k where $k = l, g$, which illustrate the various physical transport processes occurring in the fluid flow: the rate of changes of ϕ_k due to the *local acceleration* term accompanied by the *advection* term on the left hand side equivalent to the *diffusion* term ($\Gamma_{\phi_k}^k$ is designated as the diffusion coefficient) and the *source* term $S_{\phi_k}^k$ on the right hand side. In order to bring forth the common features, terms that are not shared between the equations are placed into the source terms. By setting the transport variable ϕ_k equal to 1, u_k, v_k, w_k, H_k, k_k and ε_k and selecting appropriate values for the diffusion coefficient $\Gamma_{\phi_k}^k$ and source term $S_{\phi_k}^k$, special forms of each of the partial differential equations for the continuity, momentum and energy as well as for the turbulent scalars for the two-fluid model can thus be obtained.

For subcooled boiling flow system, computational fluid dynamics which emphasizes the resolution of the physical processes through the use of digital computers proceeds by first negotiating the sub-division of the domain into a number of finite, non-overlapping sub-domains. This leads to the construction of an overlay *mesh* of *cells* (elements or control volumes) covering the whole domain. In general, the set of fundamental *mathematical* equations are required to be converted into suitable *algebraic* forms, which are then solved via suitable numerical techniques. In this sense, the original partial differential equations being considered have undergone the process of *discretisation*. Amongst the many available discretisation methods, finite volume method represents the most commonly method that is widely applied in computational fluid dynamics applications. This particular method provides the flexibility of accommodating not only meshes of regular arrangement such as structured or body-fitted meshes but also meshes of irregular structure such as unstructured meshes in order to better handle geometries having rather complex and arbitrary shapes in nature.

The cornerstone of the finite volume method is the *control volume integration*. In order to numerically solve the approximate forms of equation (18), it is convenient to consider its integral form of this generic transport equation over a finite control volume. Integration of the equation over a three-dimensional control volume ΔV yields:

$$\int_{\Delta V} \frac{\partial(\rho_k \alpha_k \phi_k)}{\partial t} dV + \int_{\Delta V} \nabla \cdot (\rho_k \alpha_k \bar{u}_k \phi_k) \, dV = \int_{\Delta V} \nabla \cdot \left[\alpha_k \Gamma_{\phi_k}^k \nabla \phi_k \right] dV + \int_{\Delta V} S_{\phi_k}^k \, dV \quad (19)$$

By applying the Gauss' divergence theorem to the volume integral of the *advection* and *diffusion* terms, equation (19) can now be expressed in terms of the elemental dA as:

$$\int_{\Delta V} \frac{\partial(\rho_k \alpha_k \phi_k)}{\partial t} dV + \int_{\Delta A} (\rho_k \alpha_k \bar{u}_k \phi_k) \cdot \bar{n} \, dA = \int_{\Delta A} \left[\alpha_k \Gamma_{\phi_k}^k \nabla \phi_k \right] \cdot \bar{n} \, dA + \int_{\Delta V} S_{\phi_k}^k \, dV \quad (20)$$

Equation (20) needs also to be further augmented with an integration over a finite time step Δt. By changing the order of integration in the time derivative terms,

$$\int_{\Delta V}\left(\int_{t}^{t+\Delta t}\frac{\partial(\rho_k\alpha_k\phi_k)}{\partial t}dt\right)dV + \int_{t}^{t+\Delta t}\left(\int_{\Delta A}(\rho_k\alpha_k\bar{u}_k\phi_k)\cdot\bar{n}\,dA\right)dt =$$
$$\int_{t}^{t+\Delta t}\left(\int_{\Delta A}\left[\alpha_k\Gamma_{\phi_k}^k\nabla\phi_k\right]\cdot\bar{n}\,dA\right)dt + \int_{t}^{t+\Delta t}\int_{\Delta V}S_{\phi_k}^k\,dV\,dt \qquad (21)$$

In essence, the finite volume method discretises the integral forms of the transport equations directly in the physical space. If the physical domain is considered to be subdivided into a number of finite contiguous control volumes, the resulting statements express the exact conservation of property ϕ_k from equation (21) for each of the control volumes. In a control volume, the bounding surface areas of the element are, in general, directly linked to the discretisation of the advection and diffusion terms. The discretised forms of these terms are:

$$\int_{\Delta A}(\rho_k\alpha_k\bar{u}_k\phi_k)\cdot\bar{n}\,dA \approx \sum_{f}(\rho_k\alpha_k\bar{u}_k\cdot\bar{n}\phi_k)_f \Delta A_f \qquad (22)$$

$$\int_{\Delta A}\left[\alpha_k\Gamma_{\phi_k}^k\nabla\phi_k\right]\cdot\bar{n}\,dA \approx \sum_{f}\left[\alpha_k\Gamma_{\phi_k}^k\nabla\phi_k\cdot\bar{n}\right]_f \Delta A_f \qquad (23)$$

where the summation in equations (22) and (23) is over the number of faces of the element and ΔA_f is the area of the face of the control volume. The source term can be subsequently approximated by:

$$\int_{\Delta V}S_{\phi_k}^k\,dV \approx S_{\phi_k}^k\Delta V \qquad (24)$$

For the time derivative term, the commonly adopted first order accurate approximation entails:

$$\int_{t}^{t+\Delta t}\frac{\partial(\rho_k\alpha_k\phi_k)}{\partial t}dt \approx \frac{(\rho_k\alpha_k\phi_k)^{n+1} - (\rho_k\alpha_k\phi_k)^n}{\Delta t} \qquad (25)$$

where Δt is the incremental time step and the superscripts n and $n+1$ denote the previous and current time levels respectively. Equation (21) can then be iteratively solved accordingly to the *fully implicit procedure* by

$$\frac{(\rho_k \alpha_k \phi_k)^{n+1} - (\rho_k \alpha_k \phi_k)^n}{\Delta t} + \left(\sum_f (\rho_k \alpha_k \vec{u}_k \cdot \vec{n} \phi_k)_f \Delta A_f \right)^{n+1} =$$
$$\left(\sum_f \left[\alpha_k \Gamma_{\phi_k}^k \nabla \phi_k \cdot \vec{n} \right]_f \Delta A_f \right)^{n+1} + \left(S_{\phi_k}^k \right)^{n+1} \Delta V \qquad (26)$$

ASSEMBLY OF DISCRETISED EQUATIONS

Depending on the particular mesh system, equation (26) can be written in a discretised form:

$$\frac{(\rho_k \alpha_k \phi_k)^{n+1} - (\rho_k \alpha_k \phi_k)^n}{\Delta t} + \left(\sum_{i=1}^{N_f} (\rho_k \alpha_k \vec{u}_k \cdot \vec{n})_i \Delta A_i \phi_{k,i} \right)^{n+1} =$$
$$\left(\sum_{i=1}^{N_f} \left[\alpha_k \Gamma_{\phi_k}^k \nabla \phi_k \cdot \vec{n} \right]_i \Delta A_i \right)^{n+1} + \left(S_{\phi_k}^k \right)^{n+1} \Delta V \qquad (27)$$

where ΔA_i is the discrete area and N_f represents the total number of faces of the control volume element. In two dimensions, a triangular element has a total number of three faces, $N_f = 3$, while a quadrilateral element has a total number of four faces, $N_f = 4$. In three dimensions, a tetradedral element has a total number of four faces, $N_f = 4$, while a hexahedral element has a total number of six faces, $N_f = 6$. For a polyhedral element, it can however accommodate substantially greater number of faces bounding the control volume.

Consider for instance the particular control volume element of which point P is taken to represent the centriod of the control volume in question, which is connected with the respective centroids of other surrounding control volumes. Equation (27) can be expressed in terms of the transport quantities at point P and surrounding nodal points with a suitable prescription of normal vectors at each control volume face and dropping the superscript $n+1$ which by default denotes the current time level as:

$$a_P^k \phi_{k,P} = \sum_{nb} a_{nb}^k \phi_{k,nb} + S_{non}^k + S_u^k \Delta V_P + \frac{(\rho_k \alpha_k \phi_k)_P^n \Delta V_P}{\Delta t} \qquad (28)$$

where

$$a_P^k = \sum_{nb} a_{nb}^k + S_P^k \Delta V_P + \sum_{i=1}^{N_f} \alpha_{k,i} F_i^k + \frac{(\rho_k \alpha_k)_P \Delta V_P}{\Delta t} \qquad (29)$$

and the added contribution due to non-orthogonality of the mesh is given by S_{non}^k, which is required to be ascertained especially for body-fitted and unstructured meshes. Note that the convective flux is given by $F_i^k = (\rho_k \vec{u}_k \cdot \vec{n})_i \Delta A_i$. For the sake of numerical treatment, the source term for the control volume in equation (27) has been treated by

$$S_{\phi_k}^k \Delta V = (S_u^k - S_P^k \phi_{k,P}) \Delta V_P \qquad (30)$$

In equation (29), a_P^k is the diagonal matrix coefficient of $\phi_{k,P}$, $\sum_{i=1}^{N_a} \alpha_{k,i} F_i^k$ are the mass imbalances over all faces of the control volume and S_P^k is the coefficient that is extracted from the treatment of the source term in order to further increase the diagonal dominance. On the basis of the definition of the Peclet number – $Pe_i^k = F_i^k / D_i^k$ – the coefficients of any neighboring nodes for any surrounding control volumes a_{nb}^k can be expressed by

$$a_{nb}^k = \alpha_{k,i} D_i^k \max\left(1 - \frac{|Pe_i^k|}{2}, 0\right) + \alpha_{k,i} \max(-F_i^k, 0) \qquad (31)$$

where D_i^k is the diffusive flux containing the diffusion coefficient $\Gamma_{\phi_k}^k$ along with the geometrical quantities of the particular element within the mesh system. In this study, the treatment of the advection term which results in the form presented in equation (31) is known as the *hybrid-differencing* which essentially retains a second order accuracy for small Peclet numbers due to central differencing but reverts to the first order upwind differencing for large Peclet numbers.

SOLUTION ALGORITHMS

The process of obtaining the computational solution of the conservation equations governing the transport of fluid and heat of multiphase flows consists of three stages. The *first* stage concerns the generation of a suitable mesh which concerns the application of various types of meshes – structured, body-fitted and unstructured meshes – to handle different multiphase flow configurations. The *second* stage involves the conversion of the partial differential equations into a system of discrete algebraic equations. The *third* stage essentially describes the array of solution algorithms required to solve the algebraic transport equations. It can be sub-divided into two categories: (i) Pressure-velocity linkage methods and (ii) Matrix solvers.

Solution algorithm based on the single phase SIMPLE (Semi-Implicit for Method Pressure-Linkage Equations) is well suited to solve the discretised macroscopic balance equations of mass, momentum and energy. Originally pioneered by Patankar and Spalding [32], this pressure correction technique is basically an iterative approach for implicit-type algorithms of steady or unsteady solutions and is centered on the basic philosophy of effectively coupling between the pressure and the velocity of which the pressure is linked to the velocity via the construction of a pressure field to guarantee conservation of mass. In this sense, the equation for the mass conservation becomes now a *kinematic* constraint on the velocity field rather than a *dynamic* equation. In the consideration of solution methods for multiphase flows, it can be demonstrated that the Inter-Phase Slip Algorithm (IPSA) and its variant Inter-Phase Slip Algorithm – Coupled (IPSA-C) are mere extensions of the well-known solution algorithm SIMPLE.

Through the IPSA method, computation of multiphase flows in the context of computational fluid dynamics can be enhanced through the use of the shared pressure approximation. This means that all phase pressures in the momentum equation are assumed to be equal, i.e. $P_k = P$. As a result, the pressure gradient term of all phases appears as a product of common pressure gradient ∇P and the respective volume fraction α_k, and accordingly forms a pressure shared by volume fraction. Because of the shared pressure approximation, a global equation for the pressure correction is sought so as to satisfy the total mass balance instead of a pressure correction for each phase. The total mass balance can be achieved by adding all the mass balances together which results in the joint equation for the conservation of mass. In addition, the IPSA-C method has been developed to alleviate the problem associated with the strong coupling between two or multiple phases. In the IPSA-C method, the performance of the pressure correction step is improved by the semi-implicit inclusion of the interfacial source terms following the idea of the SImultaneous solution of Non-linearly Coupled Equations (SINCE) method such as described in Karema and Lo [33] in order to implicitly treat the interaction of multiple phases within the interfacial source terms. For a tighter coupling between the gas and liquid phases in subcooled boiling flow, the IPSA-C method is preferred.

Provided that the pressure field is sufficiently smooth, the *standard* Rhie-Chow interpolation expressions as derived from above for different mesh systems may suffer from large errors especially in the case of rapidly changing source terms. For example, the prevalence of strong buoyancy force on different sides of a sharp interface in multiphase flows. In such a situation, this interpolation formula which involves only the evaluation of the velocities and pressure gradients by linear interpolation from cell centers may not be accurate enough, and in turn could result in significant mass imbalance in cells adjacent to the interface. To overcome the problem associated with collocated arrangement, one possible solution is to somehow mimic the staggered or mesh-oriented situation by ensuring the source term is defined primitively at cell faces, and the Rhie-Chow interpolation expressions are accordingly modified to include the effect of this term into the formula in order to appropriately evaluate the face velocities. This so-called *improved* Rhie-Chow interpolation method is adopted in this present study. More details on the description of this interpolation method can be found in Karema [34].

Efficient matrix solvers such as the Strongly Implicit Procedure (SIP) and thte combination of solver and preconditoner referred to as Incomplete Cholesky – Conjugate Gradient (ICCG) are adopted to solve the algebraic transport equations. Specifically, the ICCG matrix solver is employed to enhance the convergence of the pressure correction equation.

Chapter 4

APPLICATION OF EMPIRICAL RELATIONSHIPS TO MODELING SUBCOOLED BOILING FLOW

MECHANISTIC MODELS AND CLOSURE THROUGH VARIOUS EMPIRICAL RELATIONSHIPS

A number of mechanistic models have been developed for the prediction of wall heat flux and partitioning. Del Valle and Kenning [35] concentrated on the formulation of a mechanistic model for subcooled nucleate flow boiling by taking into consideration the bubbly dynamics at the heated wall. This model employed some of the concepts developed initially by Graham and Hendricks [36] for wall heat flux partitioning during pool nucleate boiling. The mechanistic model by Kurul and Podowski [27], adapted by Judd and Hwang [37] on the premise of a model also for wall heat flux partitioning during pool nucleate boiling, is still widely employed in the CFD commercial code CFX4.4 and ANSYS-CFX11. In the paper by Steiner et al. [38], a modified Chen-type superposition model was used to compute the wall heat flux in subcooled boiling flow. More recently, a new approach to the partitioning of the wall heat flux has been proposed by Basu et al. [39,49] and reviewed in Warrier and Dhir [41]. The fundamental idea of the model is to consider all the energy from the wall is transferred to the liquid adjacent to the heated wall. Thereafter, a fraction of the energy is transferred to the vapor bubbles by evaporation while the remainder goes into the bulk liquid.

The mechanistic model of Kurul and Podowski has been applied with much success in many CFD investigations of subcooled boling flows at low pressures. In essence, this model entails the partitioning of the wall heat flux into three heat flux components. They are: (i) Heat transferred by conduction to the superheated layer next to the wall (nucleate boiling or surface quenching), Q_q; (ii) Heat transferred by evaporation or vapour generation, Q_e; and (iii) Heat transferred by turbulent convection, Q_c. The wall heat flux partitioning can be written as

$$Q_w = Q_q + Q_e + Q_c \tag{32}$$

The surface quenching heat flux is determined through the relationship:

$$Q_q = \left[\frac{2}{\sqrt{\pi}} \sqrt{k_l \rho_l C_{pl}} \sqrt{f} \right] A_q (T_w - T_l) \tag{33}$$

where T_w is the wall temperature, A_q is the fraction of wall area subjected to cooling by quenching and f is the bubble departure frequency.

Heat flux due to vapour generation at the wall in the nucleate boiling region can simply be calculated from Bowring [42]:

$$Q_e = N_a f \left(\frac{\pi}{6} d_{bw}^3 \right) \rho_g h_{fg} \tag{34}$$

The heat flux according to the definition of local Stanton number St for turbulent convection is given as:

$$Q_c = St \rho_l C_{pl} u_l (T_w - T_l)(1 - A_q) \tag{35}$$

where u_l is the local tangential liquid velocity adjacent to the heated surface. The evaluation of the area of the heater surface influenced by bubbles A_q is given by

$$A_q = K \frac{\pi d_{bw}^2}{4} N_a \tag{36}$$

where the empirical constant K is used to account for the area of the heater surface influenced by the bubble. A value of $K = 4$ is often recommended [27]. However, Kenning and Del Valle [43] have found values ranging between 2 and 5. Judd and Hwang [37] ascertained that a lower value, $K = 1.8$, best fitted their experimental data. In most of our studies [44-46], we have incorporated a Jacob number (Ja_{sub})

based on liquid subcooling dependence as suggested by Kenning and Del Valle [43] according to

$$K = 4.8\exp\left(-\frac{Ja_{sub}}{80}\right) \tag{37}$$

To determine each respective component on the right hand side of equation (32), a bisection algorithm, an iterative procedure, is employed to evaluate the wall superheat that satisfies the applied wall heat flux. This algorithm begins with a guess of the wall superheat, and thereafter calculates each component of the heat flux. The difference between the computed total wall heat flux and the actual applied wall heat flux provides a new wall superheat estimate for the next step in iterative procedure. The iteration continues until the difference error between the applied and calculated wall heat flux falls below a prescribed criterion (i.e. $< 10^{-4}$) of the applied heat flux.

A number of studies examining bubble growth and detachment have resulted in a number of different empirical correlations for bubble departure. We focus on the selected relationships that are relevant to low-pressure subcooled boiling flow. Tolubinsky and Kostanchuk [47] proposed a simple relationship which evaluated the bubble departure as a function of the subcooling temperature as:

$$d_{bw} = \min\left[0.0006\exp\left(-\frac{T_{sub}}{45}\right), 0.00014\right] \tag{38}$$

Incidentally, Kurul and Podowski [27] also employed a similar expression to Tolubinsky and Kostanchuk [47] in calculating the bubble departure diameter. The correlation, based on Unal's experimental data [48], is:

$$d_{bw} = 0.00014 + 10^{-4} T_{sub} \tag{39}$$

On the basis of the balance between the buoyancy and surface tension forces at the heating surface, Fritz [49] proposed a correlation which includes the contact angle of the bubble, viz.,

$$d_{bw} = 0.0208\theta\sqrt{\frac{\sigma}{g(\rho_l - \rho_g)}} \tag{40}$$

where θ is the contact angle. The above expression was modified by Kocamustafaogullari and Ishii [50] for low pressure as

$$d_{bw} = 2.5 \times 10^{-5} \left(\frac{\rho_l - \rho_g}{\rho_g} \right) \theta \sqrt{\frac{\sigma}{g(\rho_l - \rho_g)}} \qquad (41)$$

A more comprehensive correlation proposed by Unal [48] which included not only the effect of subcooling but also on the convection velocity and heater wall properties is given by:

$$d_{bw} = \frac{2.42 \times 10^{-5} \, p^{0.709} \, a}{\sqrt{b\Phi}} \qquad (42)$$

where

$$a = \frac{(Q_w - hT_{sub})^{1/3} k_l}{2C^{1/3} h_{fg} \sqrt{(\pi k_l / \rho_l c_{pl} \rho_g)}} \sqrt{\frac{k_w \rho_w c_{pw}}{k_l \rho_l c_{pl}}} \; ; \; b = \frac{T_{sub}}{2\left[1 - (\rho_g / \rho_l)\right]}$$

$$C = \frac{h_{fg} \mu_l \left[c_{pl} / (0.013 h_{fg} \, \text{Pr}^{1.7}) \right]^3}{\sqrt{\frac{\sigma}{g(\rho_l - \rho_g)}}} \; ;$$

$$\Phi = \begin{cases} \left(\dfrac{u_l}{0.61} \right)^{0.47} & \text{for } u_l \geq 0.61 \, m/s \\ 1.0 & \text{for } u_l < 0.61 \, m/s \end{cases}$$

The stated range for this correlation is

Pressure $0.1 < p < 17.7$ MPa
Wall heat flux $0.47 < Q_w < 10.64$ MW / m^2
Liquid velocity $0.08 < u_l < 9.15$ m/s
Liquid subcooling $3.0 < T_{sub} < 86 \, ^\circ C$

For the bubble departure frequency, most correlations have been derived from the consideration of the bubble departure diameter. Cole's correlation [51] which was derived assuming a balance between buoyancy and drag (drag coefficient constant) for pool nucleate boiling is a popular expression. It is used as a default expression in the CFD commercial code CFX4.4 in the form of:

$$fd_{bw}^{0.5} = \sqrt{\frac{4g(\rho_l - \rho_g)}{3\rho_l}} \qquad (43)$$

In the hydrodynamic region, Ivey [52] has suggested a bubble frequency relationship, which can be used for coalesced bubbles. The correlation is simply

$$fd_{bw}^{0.5} = 0.9\sqrt{g} \qquad (44)$$

A slightly more complicated expression by Stephan [53], which included the effect of surface tension, may also be employed for the low-pressure subcooled boiling flow. It is given as:

$$fd_{bw} = \frac{1}{\pi}\sqrt{\frac{g}{2}\left(d_{bw} + \frac{4\sigma}{\rho g d_{bw}}\right)} \qquad (45)$$

Similar expressions of the bubble frequency correlation that include the surface tension effect in the form of equation (45) are also noted. Peebles and Garber [54] observed the velocity rise in a gravitational field as

$$V_b = 1.18\left[\frac{\sigma g(\rho_l - \rho_g)}{\rho_l^2}\right]^{0.25} \qquad (46)$$

On the basis of equation (32), Jakob [55] proposed a bubble frequency by assuming the waiting time to be equivalent to the growth time thus resulting in

$$fd_{bw} = 0.59\left[\frac{\sigma g(\rho_l - \rho_g)}{\rho_l^2}\right]^{0.25} \qquad (47)$$

The Lemmert and Chwala's [56] active nucleation site density relationship is commonly used in the wall heat flux partitioning model of which has been correlated based on Del Valle and Kenning [35] data. It can be determined simply from the local wall superheat as

$$N_a = \left[m \left(T_{sat} - T_w \right) \right]^n \qquad (48)$$

According to Kurul and Podowski [27], the values of m and n are 210 and 1.805 respectively. Recently, Končar et al. [57] demonstrated that better predictions were obtained by reducing the value of m to 185. Other relationships of the active nucleation site density that may also be applicable to low-pressure subcooled boiling flow are subsequently described. Kocamustafaogullari and Ishii [50] correlated existing active nucleation site density data by means of parametric study. They have assumed that the active nucleation site density in pool boiling by both surface conditions and thermo-physical properties of the fluid. They also postulated that the active nucleation site density developed for pool boiling could be used in forced convective system by the use of an effective superheat rather than the actual wall superheat. The active nucleation site density, N_a, can be expressed as:

$$N_a = \frac{1}{d_{bw}^2} \left[\frac{2\sigma T_{sat}}{\Delta T_{eff} \rho_g h_{fg}} \right]^{-4.4} f(\rho^*) \qquad (49)$$

where $\rho^* = (\rho_l - \rho_g)/\rho_g$ and the function $f(\rho^*)$ is a known function of a density ratio described by: $f(\rho^*) = 2.157 \times 10^{-7} \rho^{*-3.2} (1 + 0.0049 \rho^*)^{4.13}$. ΔT_{eff} represents the effective superheat, which is given by $\Delta T_{eff} = S \Delta T_w$ where $\Delta T_w = T_{sat} - T_w$ and S is the suppression factor. Basu et al. [58] proposed an alternative empirical correlation that included the effect of contact angle θ on the active nucleation site density, which is given by:

$$\begin{aligned} N_a &= 0.34 \times 10^4 (1 - \cos\theta) \Delta T_w^{2.0} & \Delta T_{ONB} < \Delta T_w < 15K \\ N_a &= 3.4 \times 10^{-1} (1 - \cos\theta) \Delta T_w^{5.3} & 15K \leq \Delta T_w \end{aligned} \qquad (50)$$

Very recently, Hibiki and Ishii [59] modeled the active nucleation site density relationship mechanistically by the knowledge of the size and cone angle distributions of cavities. In accordance with Basu et al. [58] correlation, they have also formulated the nucleation site density as a function of contact angle. The correlation is given by:

$$N_a = 4.72 \times 10^5 \left\{1 - \exp\left(-\frac{\theta^2}{4.17}\right)\right\} \left[\exp\left\{2.5 \times 10^{-6} f(\rho^+) \frac{\Delta T_w \rho_g h_{fg}}{2\sigma T_{sat}}\right\} - 1\right] \quad (51)$$

where $\rho^+ = \log_{10}\left((\rho_l - \rho_g)/\rho_g\right)$ and the function $f(\rho^+)$ is a function described by: $f(\rho^+) = -0.01064 + 0.48246\rho^+ - 0.22712\rho^{+^2} + 0.05468\rho^{+^3}$.

MODEL APPLICATIONS

The three most important parameters in the wall heat flux partitioning model are the bubble departure diameter d_{bw}, bubble departure frequency f and the active nucleation site density N_a. It is recognized that not one single dedicated correlation for each of the parameters is expected to be universally valid for all types of flow conditions. A survey in the literature has revealed that the mechanistic model of Kurul and Podowski is a widely used heat transfer closure model for modeling the thermal-hydraulics in nuclear reactor geometries with different correlations of d_{bw}, f and N_a. Some pertinent investigations on the application of various empirical relationships for subcooled boiling flow are presented below.

Anglart and Nylund [22] performed a subcooled boiling flow investigation to predict the void fraction distribution in two-phase bubbly flows in fuel rod bundles. Special attention was devoted into the phenomena which govern the lateral void fraction distribution in a channel. The two-fluid boiling model with the closure of appropriate relationships for the interfacial mass, momentum and energy terms and the standard k-ε turbulence model with the additional consideration of Sato's bubble-induced turbulent viscosity was employed. Assuming spherical bubbles, the interfacial area concentration can be defined as:

$$a_{if} = \frac{6\alpha_g}{D_s} \quad (52)$$

Equation (52) requires the evaluation of the local distribution of the Sauter bubble diameter D_s. In Anglart and Nylund [22], the bubble diameter has been modeled as a function of the local water subcooling T_{sub} from the following equation:

$$D_s = \begin{cases} D_{B1} = 0.15 \text{ mm} & \text{for } T_{sub} > T_{sub,1} = 13.5 \text{ K} \\ \dfrac{D_{B1}\left(T_{sub} - T_{sub,2}\right) + D_{B2}\left(T_{sub,1} - T_{sub}\right)}{\left(T_{sub,1} - T_{sub,2}\right)} & \text{for } T_{sub,2} \leq T_{sub} < T_{sub,1} \\ D_{B2} = 1.5 \text{ mm} & \text{for } T_{sub} \leq T_{sub,2} = 0 \text{ K} \end{cases} \quad (53)$$

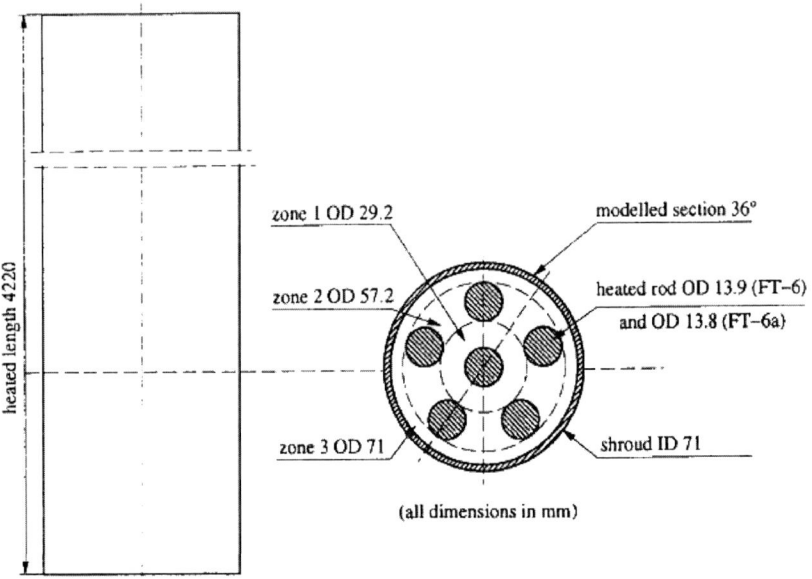

Figure 2. A schematic illustration of the test section with six rods (after Anglart and Nylund [22]).

For the wall heat flux partitioning model, closure was achieved through correlation of equation (39) for the bubble departure diameter d_{bw}, Cole's relationship for the bubble departure frequency f and the Lemnart and Chwala's expression with $m = 210$ and $n = 1.805$ for the active nucleation site density N_a. Void fraction measurements were performed by Nylund et al. [60] in the six-rod fuel bundles as shown in Figure 2.

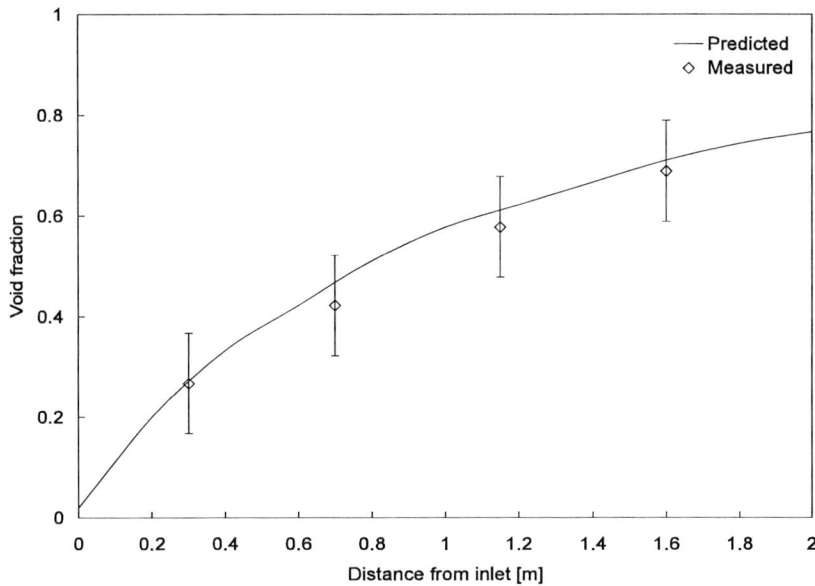

Zone 1.

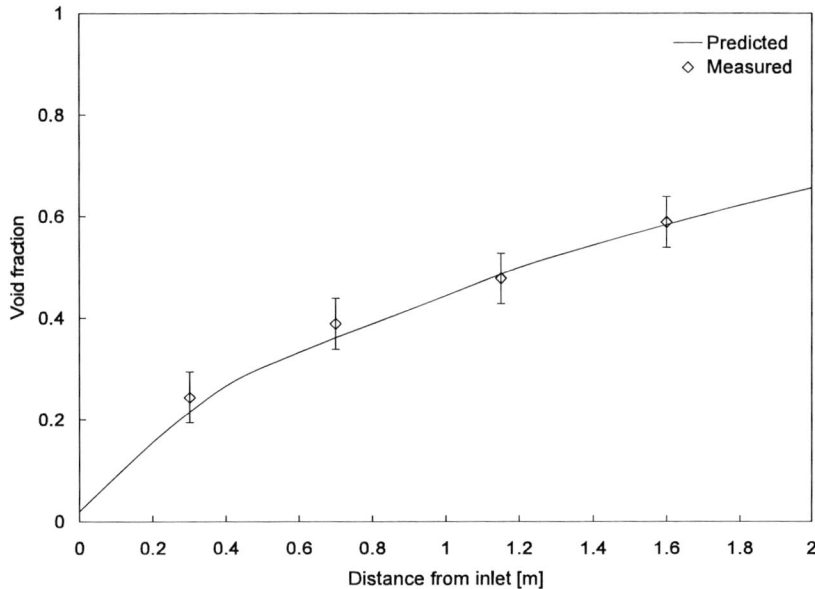

Zone 2.

Figure 3. Axial void distributions in test section (after Anglart and Nylund [22]).

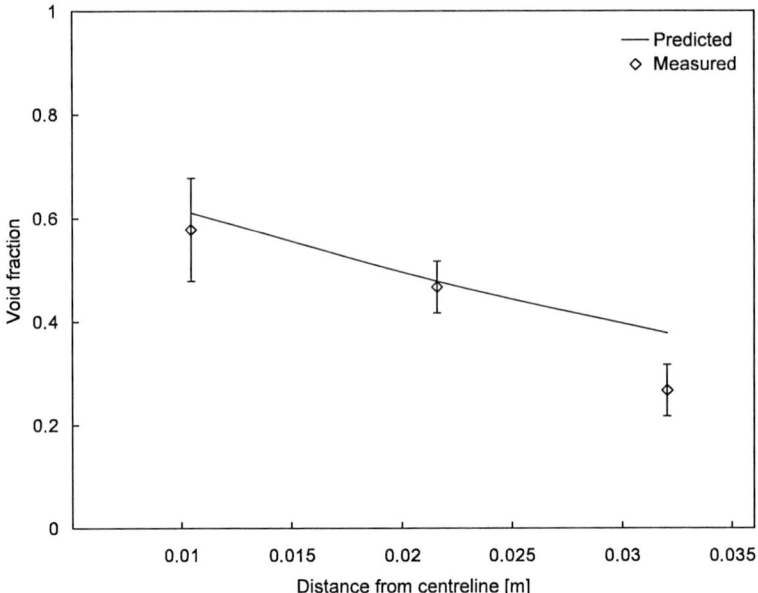

1148 mm from inlet.

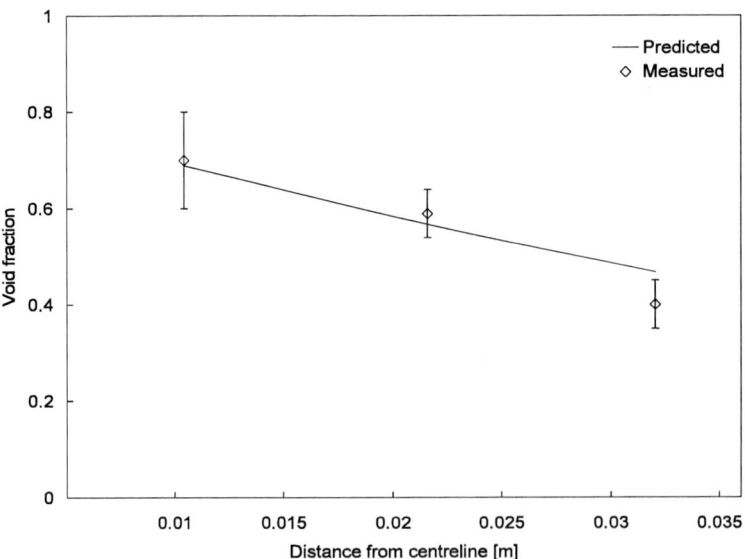

1598 mm from inlet.

Figure 4. Lateral void distributions in test section (after Anglart and Nylund [22]).

Numerical calculations were performed for one-tenth of the cross-section, utilizing the symmetry property of the bundle geometry. Comparisons of the some results of the axial and lateral distributions of the computed and measured void fraction are illustrated in figures 3 and 4. Satisfactory agreement that had been achieved between the predicted and measured void fractions at specific locations of the test section clearly highlighted the model's potential of resolving more complex fuel bundle geometries.

On the basis of the parametric study focusing on different closure empirical relationships that were carried out in our early work [44], we have established the use of Unal's bubble departure relationship (equation (42)) as the preferred model for bubble departure d_{bw}. We further advocated the use of a more sophisticated expression proposed by Zeitoun and Shoukri [61,62] for evaluating the Sauter bubble diameter D_s instead of the linear functional expression of equation (53), which is given by:

$$\frac{D_s}{\sqrt{\sigma/g\Delta\rho}} = \frac{0.0683(\rho_l/\rho_g)^{1.326}}{Re^{0.324}\left(Ja_{sub} + \frac{149.2(\rho_l/\rho_g)^{1.326}}{Bo^{0.487}Re^{1.6}}\right)} \quad (54)$$

where Re is the flow Reynolds number, Bo is the boiling number and Ja_{sub} is the Jakob number based on the liquid subcooling. To close the wall heat flux partitioning model, Cole's relationship for the bubble departure frequency f and the Lemnart and Chwala's expression with $m = 210$ and $n = 1.805$ for the active nucleation site density N_a were employed as in Anglart and Nylund [22].

Numerical predictions of the above model were examined for a range of experimental data covering a range of heat flux, subcooling and flow conditions at low pressures. For the purpose of illustration, results of axial void fraction distributions compared against experimental measurements of Donveski and Shoukri [63] and Dimmick and Selander [64] are presented herein. The data of Donveski and Shoukri [63] were obtained for a vertical 306 mm long concentric annular test section with an outer diameter of 25.4 mm. Dimmick and Selander's [64] data were obtained for subcooled boiling flow inside a tube of 12.29 mm inside diameter. These configurations are typical geometries that can be found in nuclear applications. Figures 5 and 6 illustrate the comparison between the measured and predicted void fraction. In all the cases, the agreement was found to be generally very good.

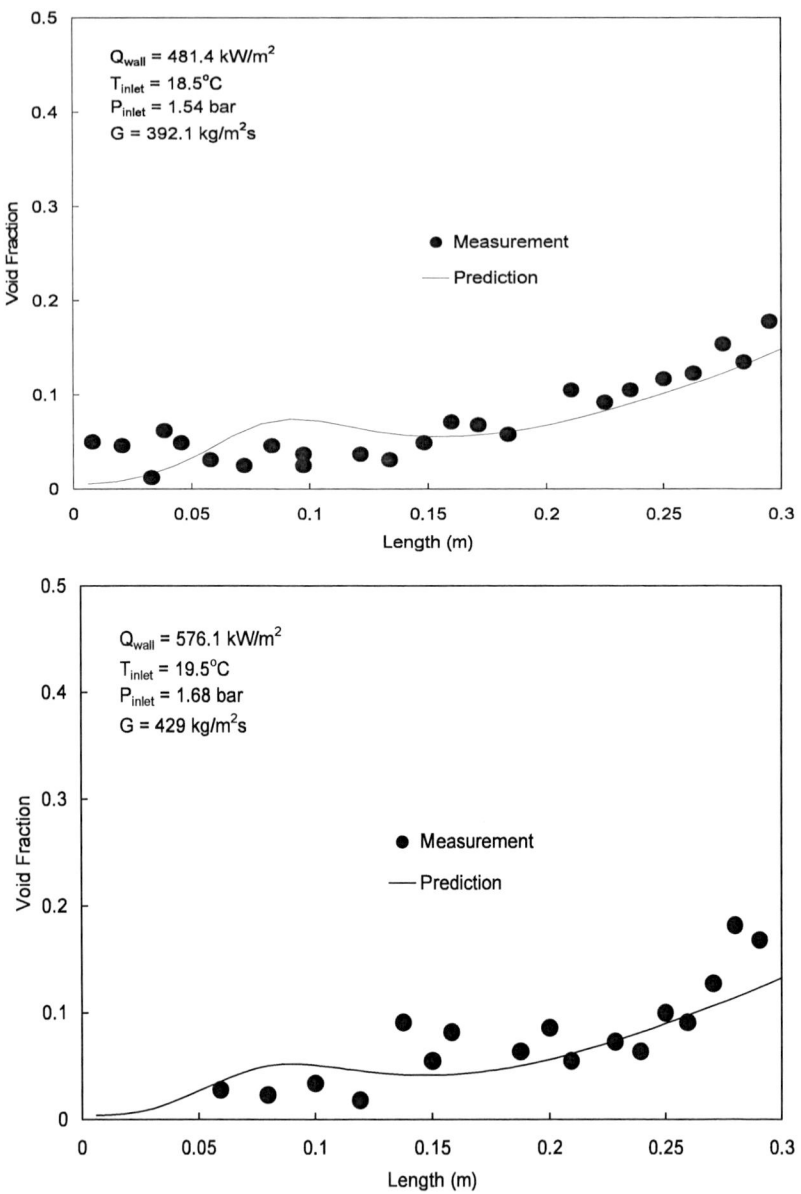

Figure 5. Comparison of predicted void fraction against Donveski and Shoukri's [63] data.

Further applications of the above boiling model in comparing measurements of a low-pressure subcooled boiling annular carried out by Yun et al. [65] and Lee et al. [66] revealed significant loss, predominantly in the radial prediction of the Sauter bubble diameter [67]. Numerical calculations were compared against measurements performed for a test channel of a 2.376 m long vertical concentric annulus with a heated inner tube. The inner tube of 19 mm in outer diameter was composed of a heated section of 1.67 m in length and two unheated sections. A transparent glass tube placed in the vicinity of the measuring plane at 1.61 m allowed visual observations of the boiling flow. Comparison of the bubble size distribution evaluated from the model employing the Zeitoun and Shoukri's relationship as described in equation (54) and calculations based on different reference diameters for the linear bubble function based on equation (53) against experimental measured data of different heat fluxes, inlet subcooling temperatures and flow conditions are shown in Figure 7. We have assumed for the case of "Linear 1" the local bubble diameters were evaluated between D_{B1} = 0.15 mm and D_{B2} = 4.0 mm while for the case of "Linear 2" they are determined between D_{B1} = 0.15 mm and D_{B2} = 7.0 mm. We further assumed that both the reference diameters corresponded to identical reference subcooling temperatures of $T_{sub,1}$ = 13.0 K and $T_{sub,2}$ = −5 K. The closure for the wall heat flux partitioning model was achieved through Kocamustafaogullari and Ishii's modified expression of Fritz for the bubble departure diameter d_{bw}, Cole's relationship for the bubble departure frequency f and the Lemnart and Chwala's expression with m = 210 and n = 1.805 for the active nucleation site density N_a. As demonstrated in Figure 7, the locally predicted bubble diameter distributions determined through the empirical relationships were not consistent with those measured during the experiments. The failure of model was predominantly attributed by the absence of the mechanistic behavior of bubble coalescence and collapse due to condensation not appropriately accounted in the model. Evidently, the bubble size distribution was not strictly dependent on only the local subcooling effect.

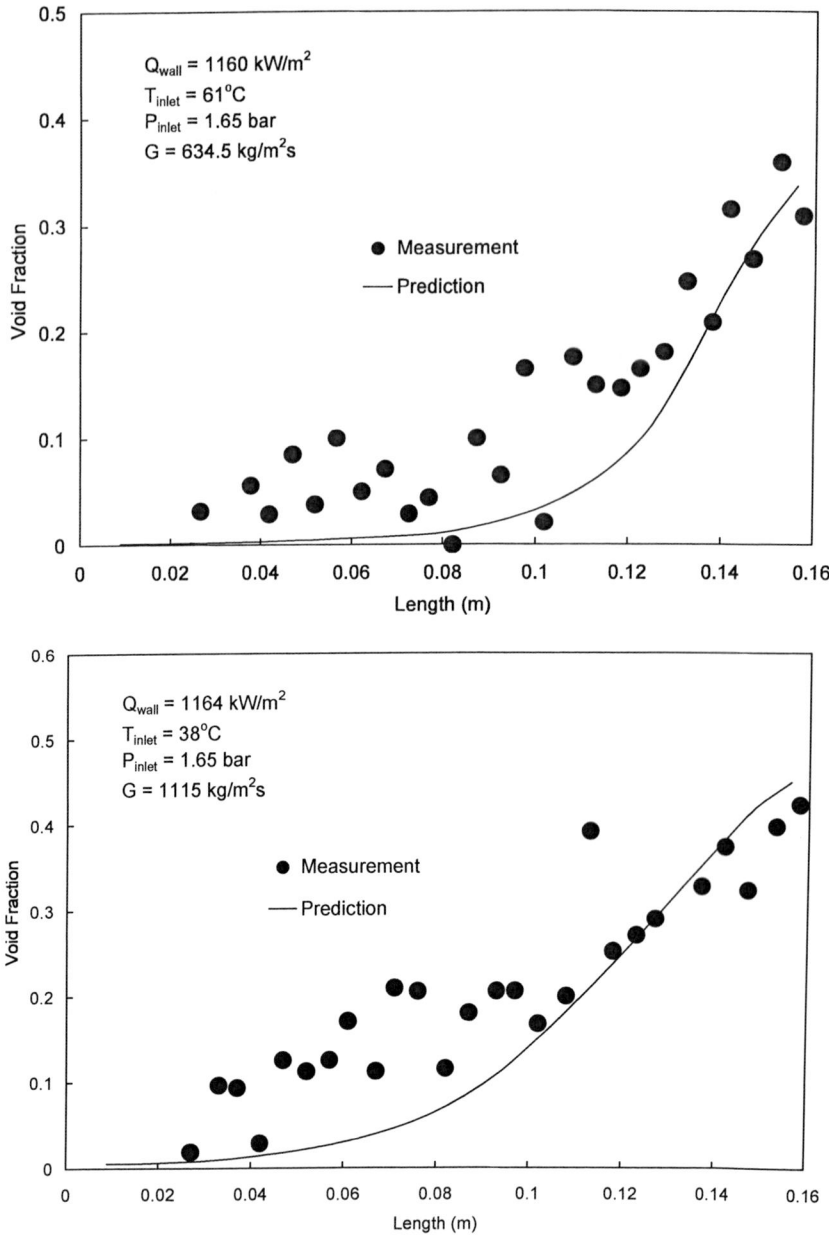

Figure 6. Comparison of predicted void fraction against Dimmick and Selander's [64].

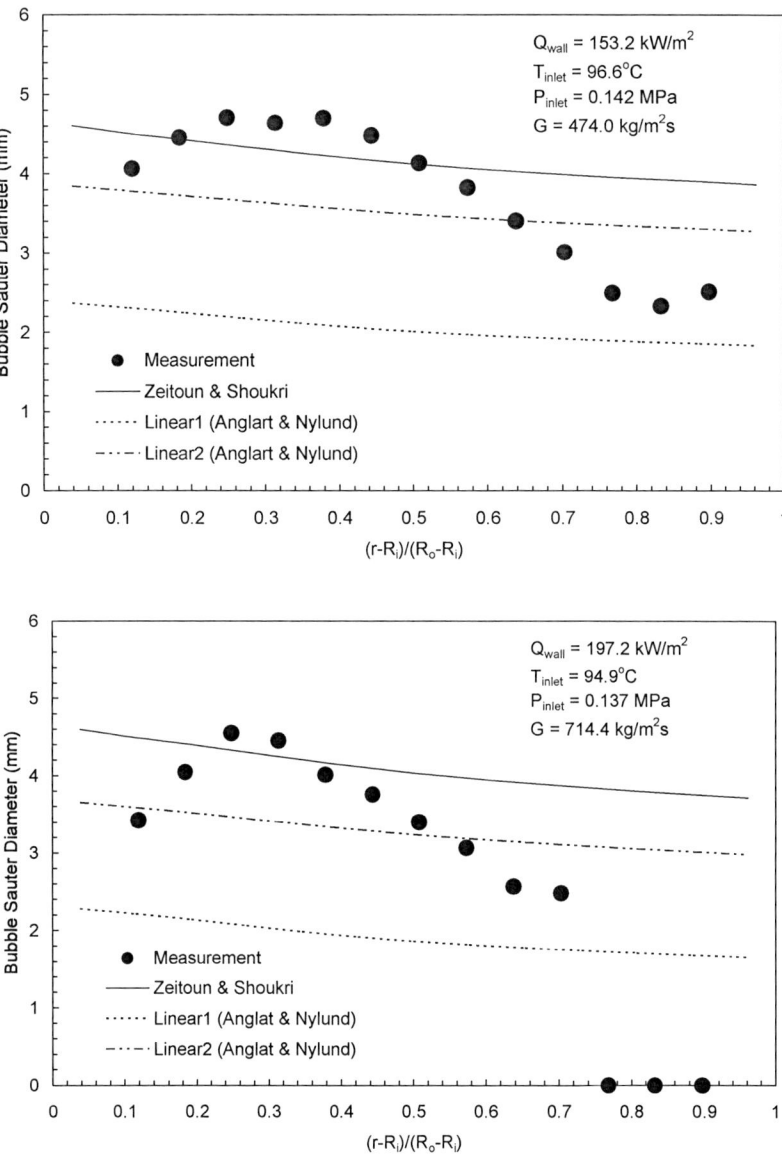

Figure 7. Continued on next page.

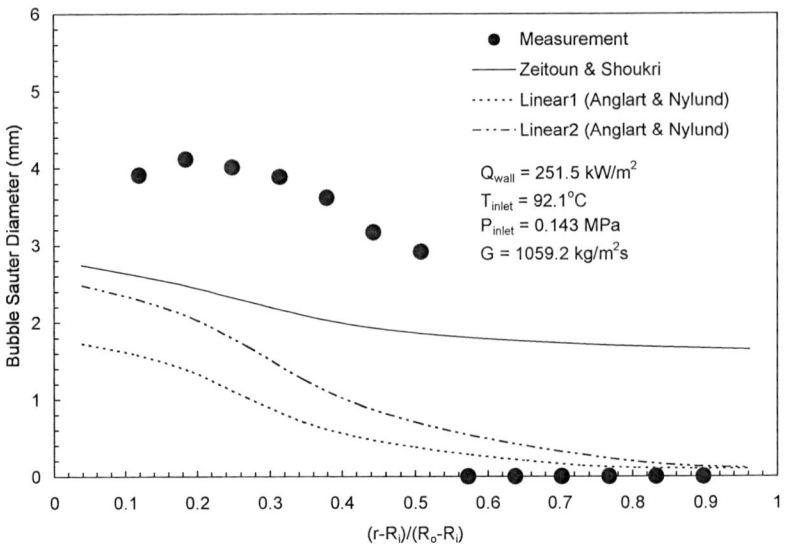

Figure 7. Comparison of local mean radial profiles of Sauter bubble diameter between predictions of empirical bubble diameter and measurements at the measuring plane.

Similar numerical investigations on the evolution of cross-sectional distribution of the local Sauter bubble diameter were also performed in Končar et al. [57]. A simple model to determine the radial distribution of the Sauter bubble diameter was proposed. The shift of the maximum bubble size away from the heated wall was modeled by a linear evolution in the radial direction as:

$$D_s^* = \min\left(d_{bw} + y_w, d_{b,\max}\right) \quad (55)$$

where y_w is the radial distance from the near-wall cell center. The maximum allowed bubble diameter $d_{b,\max}$ was prescribed as $2d_{bw}$. Multiplying equation (55) by a relation describing condensation in the subcooled flow field yields the relationship of the local Sauter bubble diameter as:

$$D_s = D_s^* \exp\left(-\frac{T_{sub} - T_{sub,w}}{2 \cdot T_{sub,w}}\right) \quad (56)$$

where $T_{sub,w}$ represents the local subcooling in the near-wall cell. To evaluate the bubble departure diameter d_{bw}, Končar et al. [57] also applied the correlation proposed by Unal [47] as was applied in our early work [44]. The correlation was nevertheless modified by adding a coefficient C_{bw} to describe large bubbles at low-pressure conditions, viz.,

$$d_{bw} = C_{bw} \frac{2.42 \times 10^{-5} p^{0.709} a}{\sqrt{b\Phi}} \qquad (57)$$

The default value of the coefficient C_{bw} was set to 2 but they have however found the need to adjust the coefficient for their numerical predictions in some experimental cases.

Of particular significance, comparison of their model predictions was made against the measurements of Lee et al. [66]. The flow conditions of the four experimental cases considered are shown in table 1. In d_{bw}, the default value of C_{bw} was used. Predicted radial distributions of the void fraction for cases stipulated in table 1 are illustrated in Figure 8. It is noted that in all the figures presented, the dimensionless parameter $(r-R_i)/(R_o-R_i) = 1$ indicates the inner surface of the unheated flow channel wall while $(r-R_i)/(R_o-R_i) = 0$ indicates the surface of the heating rod in the annulus channel. Although the measurements of local void fraction were only performed at one axial location ($z = 1.61$ m), simulations at other locations were presented to demonstrate the axial evolution of the subcooled boiling flow. Satisfactory agreement between predictions and measurement was achieved except for Case 2 where the local void fraction was significantly under-predicted. It should be noted that no comparison of their model was made especially on their predicted local Sauter bubble diameter distribution through equation (56) against local measurements.

In the recent paper by Krepper et al. [68], the mechanistic model of Kurul and Podowski [27] through closure relationships using equations (38), (43) and (48) and determination of local bubble size through linear equation (53) were investigated for their capability to contribute towards fuel assembly design. Despite the heat transfer closure achieved through these relationships, the application of CFD with the subcooled boiling model for the simulation of a hot fuel channel was shown to at least provide some means of better evaluating different geometrical designs of the spacer grids and assessing the integrity of fuel rods within a fuel assembly. They have however indicated that further improvements of the microscopic models are necessary and the model for the bubble size at departure has to be replaced by a more mechanistic model. The simulation of the bubble size in the bulk subcooled liquid has to also consider the prevalence of coalescence and fragmentation of bubbles.

Table 1. Experimental conditions of subcooled boiling flow taken from Lee et al. [66]

Experiment No.	Pinlet (MPa)	Qw (kW/m2)	G (kg/m2 s)	Tinlet (oC)	Tsub,inlet (oC)
1	0.115	169.76	478.14	83.9	19.6
2	0.121	232.59	718.16	84.0	21.2
3	0.130	114.78	476.96	95.6	11.5
4	0.125	139.08	715.17	93.9	12.0

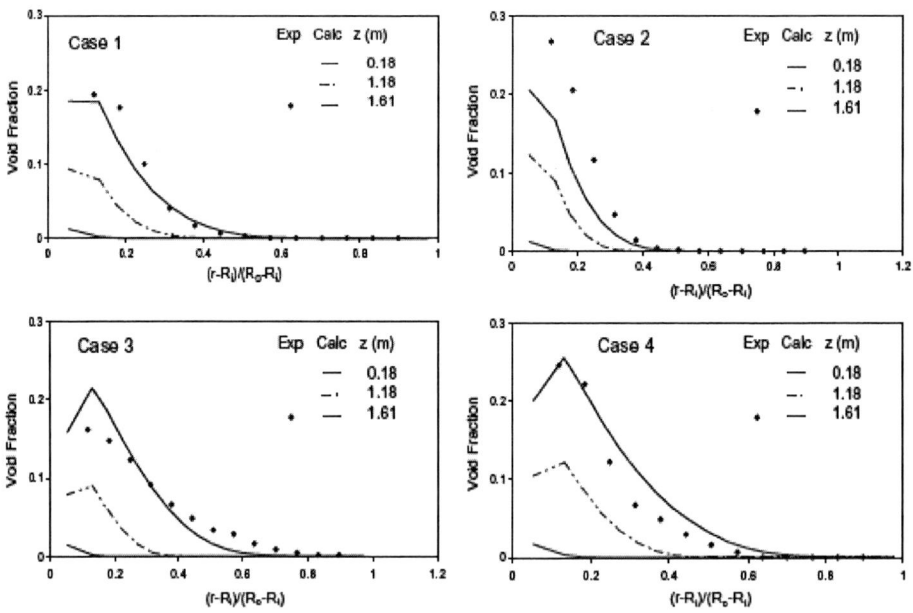

Figure 8. Comparison of predicted radial distributions of void fraction against measurements of Lee et al. [67] (after Končar et al. [57]).

Chapter 5

PHENOMENOLOGICAL OBSERVATIONS AND POPULATION BALANCE MODEL

Empirical correlations as exemplified in the previous section can at best only provide a macroscopic description of the boiling phenomenon. It is therefore not surprising that these relationships are unable to adequately represent some of the many important complex mechanistic behaviors of bubbles such as bubble coalescence and condensation (microscopic in nature) that have been observed in experiments. The importance to appropriately consider these behaviors has been clearly demonstrated by the incorrect prediction of the local Sauter bubble diameter distributions as shown in Figure 7.

As observed in experiments by Lee et al. [66] and Tu et al. [67], high-speed photography in the vicinity of the heated wall as shown in Figure 9 clearly confirmed the presence of large bubbles away from the heated wall. More importantly, the bottom vapor bubble at upstream can be seen to be sliding along the heated surface before being impeded by the downstream bubble (see Figure 9(a)). Such observation was also confirmed by experiments performed by Bonjour and Lallemand [69] and Prodanovic et al. [70], which clearly indicated the presence of bubbles sliding shortly after being detached from the heated cervices before lifting into the liquid core. In their experiments, they have also seen that these upstream bubbles traveling closely to the heated wall have the tendency of significantly colliding with any detached bubbles downstream and subsequently forming bigger bubbles due to bubble merging or coalescence of adjacent bubbles. The supposition of larger bubbles being present due to bubble coalescence was evidenced away from the heated wall as shown in Figure 9(b). Focusing on the local bubble mushroom region as depicted in Figure 9(b) and tracking its development through time, the region increased in size along the heated wall downstream, confirming the significant coalescence of bubbles. On the other hand, very few bubbles were present in the bulk subcooled liquid further away from the heated wall and near the unheated wall. In this flow region, the effect of condensation gradually caused the bubbles to decrease in size due to the subcooling temperature of the liquid as they migrated towards the opposite end of the unheated wall of the flow channel. This was further confirmed by experimental observations of Gopinath et al. [71] (see Figure 10), which illustrated a bubble gradually being condensed in a subcooled liquid away from the heated surface.

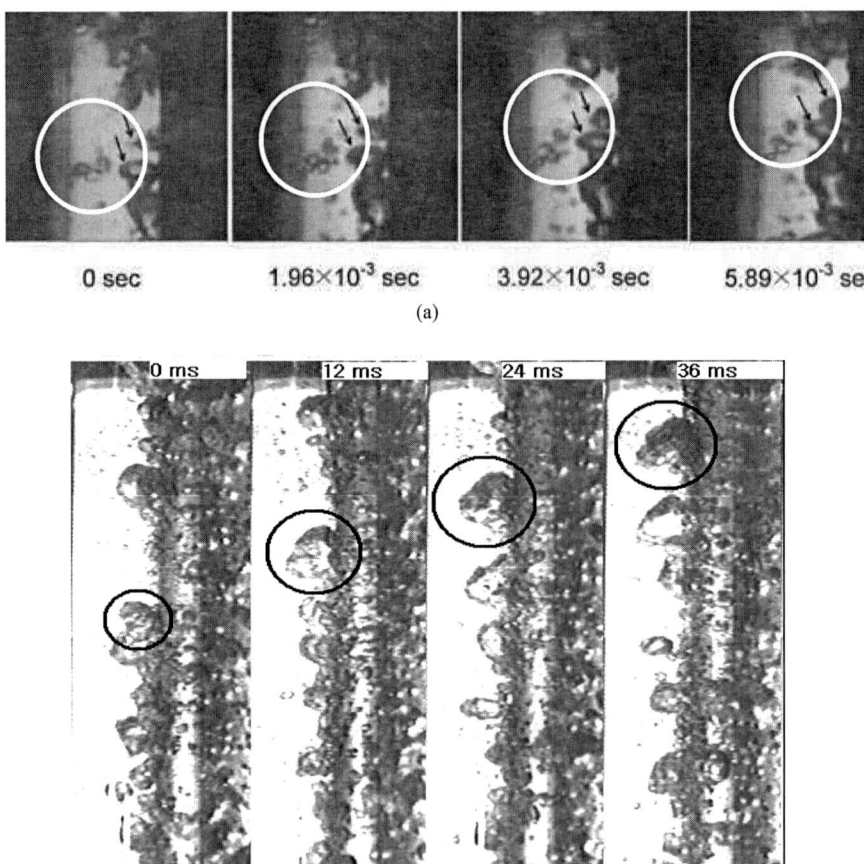

Figure 9. Experimental observations: (a) bubble sliding and collision with downstream bubble and (b) significant bubble coalescence as indicated by the bubble mushroom region near the heated wall of the annular flow channel (after Lee et al. [66] and Tu et al. [67]).

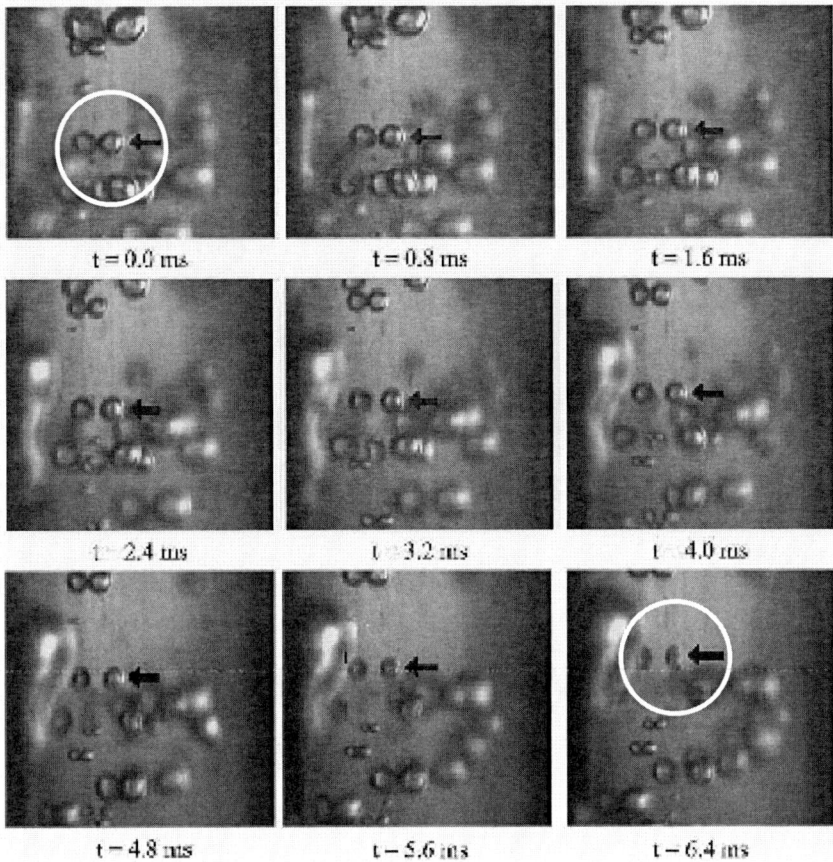

Figure 10. Bubble undergoing condensation in the bulk subcooled liquid (after Gopinath et al. [71]).

On the basis of the important phenomenological observations of the complex mechanistic behaviors of bubble coalescence near the wall and bubble collapse due to condensation in the single phase fluid flow as observed above, the population balance model which can accurately account for the coalescence and breakage characteristics of bubbles presents one of the many viable approaches in predicting the non-uniform bubble distribution in subcooled boiling flow. Phenomenological models developed by Luo and Svendsen [72] and Prince and Blanch [73] and have allowed detailed description of the mechanisms for break-up and coalescence of intermittent bubbles, which are illustrated in Figure 11.

For the consideration of the fragmentation of bubbles, the model developed by Luo and Svendsen [72] is employed for the break-up of bubbles in turbulent dispersions. In this model, binary break-up of the bubbles is assumed and the model is based on the theories of isotropic turbulence. The break-up rate of bubbles of volume v_j into volume sizes of v_i can be obtained as:

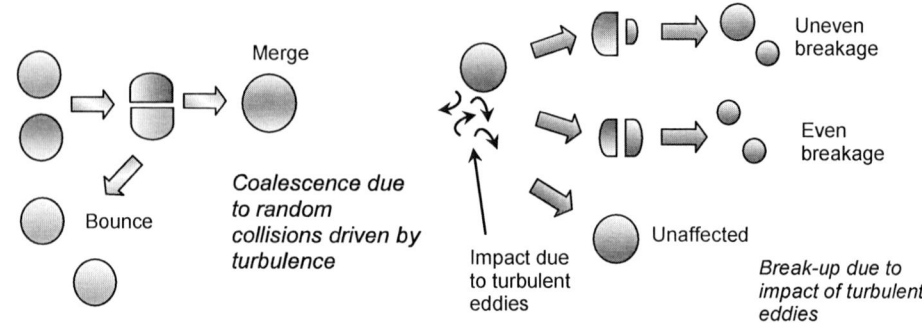

Figure 11. Bubbles undergoing coalescence and breakage.

$$\frac{\Omega(v_j : v_i)}{(1-\alpha_g)n_j} = C\left(\frac{\varepsilon}{d_j^2}\right)^{1/3} \int_{\xi_{min}}^{1} \frac{(1+\xi)^2}{\xi^{11/3}} \exp\left(-\frac{12c_f\sigma}{\beta\rho_l\varepsilon^{2/3}d_j^{5/3}\xi^{11/3}}\right) d\xi \qquad (58)$$

where $\xi = \lambda/d_j$ is the size ratio between an eddy and a particle in the inertial sub-range and consequently $\xi_{min} = \lambda_{min}/d_j$; and C and β are determined respectively from fundamental consideration of drops or bubbles breakage in turbulent dispersion systems to be 0.923 and 2.0. The variable c_f denotes the increase coefficient of surface area: $c_f = \left[f_{BV}^{2/3} + (1-f_{BV})^{2/3} - 1\right]$ where f_{BV} is the breakage volume fraction.

For the merging of bubbles, the coalescence of two bubbles is assumed to occur in three stages. The first stage involves the bubbles colliding thereby trapping a small amount of liquid between them. This liquid film then drains until it reaches a critical thickness and the last stage features the rupturing of the liquid film subsequently causing the bubbles to coalesce. The collisions between bubbles may be caused by turbulence, buoyancy and laminar shear. Only random collision driven by turbulence is considered in the present model. The coalescence rate considering turbulent collision taken from Prince and Blanch [73] can be expressed as:

$$\chi_{ij} = \frac{\pi}{4}\left[d_i + d_j\right]^2 \left(u_{ti}^2 + u_{tj}^2\right)^{0.5} \exp\left(-\frac{t_{ij}}{\tau_{ij}}\right) \tag{59}$$

where τ_{ij} is the contact time for two bubbles given by $(d_{ij}/2)^{2/3}/\varepsilon^{1/3}$ and t_{ij} is the time required for two bubbles to coalesce having diameter d_i and d_j estimated to be $\{(d_{ij}/2)^3 \rho_l/16\sigma\}^{0.5}\ln(h_0/h_f)$. The equivalent diameter d_{ij} is calculated as suggested by Chesters and Hoffman [74]: $d_{ij} = (2/d_i + 2/d_j)^{-1}$. According to Prince and Blanch [73], for air-water systems, experiments have determined h_0, initial film thickness and, h_f, critical film thickness at which rupture occurs to be 1×10^{-4} m and 1×10^{-8} m respectively. The turbulent velocity u_t in the inertial subrange of isotropic turbulence by Rotta [75] is given by:

$$u_t = 1.4\varepsilon^{1/3} d^{1/3} \tag{60}$$

POPULATION BALANCE APPROACH BASED ON THE CLASS METHOD

The foundation development of the population balance equation stems from the consideration of the Boltzman equation. Such equation is generally expressed in an *integrodifferential* form describing the particle size distribution:

$$\frac{\partial f(\xi,\vec{r},t)}{\partial t} + \nabla \cdot \left(u(\xi,\vec{r},t) f(\xi,\vec{r},t)\right) = \\ \frac{1}{2}\int_0^\xi a(\xi-\xi',\xi')f(\xi-\xi',t)f(\xi',t)d\xi' - \\ \int_0^\infty a(\xi-\xi',\xi')f(\xi,t)f(\xi',t)d\xi' + \\ \int_\xi^\infty \gamma(\xi')b(\xi')p(\xi/\xi')f(\xi',t)ds - b(\xi)f(\xi,t) \tag{61}$$

where $f(\xi,\vec{r},t)$ is the particle size distribution dependent on the internal space ξ, whose components could be characteristics dimensions, surface area, volume and so on. $\vec{r}$ and t are the external variables representing the spatial position vector and physical time in external coordinate respectively. $u(\xi,\vec{r},t)$ is velocity vector in external space. On the right hand side of equation (61), the first and second terms denote birth and death rate of particle of space ξ due to merging processes such as: coalescence or agglomeration processes; the third and fourth terms account for the birth and death rate caused by the breakage processes respectively. $a(\xi,\xi')$ is the coalescence or agglomeration rate between particles of size ξ and ξ'. Conversely, $b(\xi)$ is the breakage rate of particles of size ξ. $\gamma(\xi')$ is the number of fragments/daughter particles generated from the breakage of a particle of size ξ' and $p(\xi/\xi')$ represents the probability density function for a particle of size ξ to be generated by breakage of a particle of size ξ'.

In the method of discrete classes, which is known as the class method, the continuous size range of particles is discretized into a series number of discrete size classes. For each class, a scalar (number density of particles) equation is solved to accommodate the population changes caused by intra/inter-group particle coalescence and breakage. The particle size distribution is thereby approximated as follow:

$$f(\xi,t) \approx \sum_{i=1}^{N} n_i \delta(\xi - x_i) \qquad (62)$$

where n_i represents the number density or weight of the ith class consists of all particles per unit volume with a pivot size or abscissa, x_i of which the groups (or abscissas) of class methods are fixed and aligned continuously in the state space. A graphical representation of the class method in approximating the particle size distribution is depicted in Figure 12. The solution due to the class method has been found to be independent of the resolution of the internal coordinate if sufficient number of classes are adopted. Computationally, as the number of transport equations depends on the number of group adopted, it requires more computational time and resources to achieve stable and accurate numerical predictions.

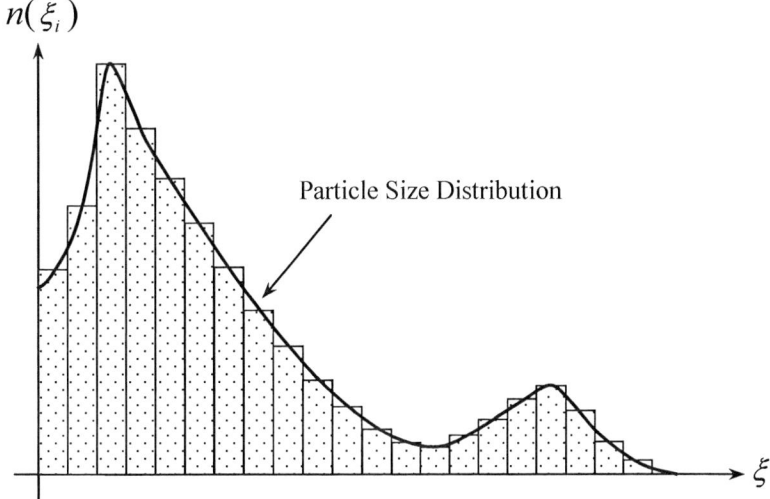

Figure 12. Graphical illustration of class method.

According to Fleischer et al. [76], the number density equation can be expressed in the form:

$$\frac{\partial n(V,\vec{r},t)}{\partial t} + \nabla \cdot \left(\vec{u}_g n(V,\vec{r},t)\right) = G(V,\vec{r},t) \tag{63}$$

where $n(V, \vec{r}, t)$ is the bubble number density distribution per unit mixture of which is a function of the spatial range $\vec{r}$ for a given time t and volume V. On the right hand side, the term $G(V, \vec{r}, t)$ contains the bubble source/sink rates per unit mixture volume due to the bubble interactions such as coalescence and break-up as well as phase change, which will be detailed below. For the case of nucleate boiling and condensation in a subcooled boiling flow, the phase change term includes the rate of change of bubble population with specific volumes. To effectively employed the number density transport equation in the context of CFD, Pohorecki et al. [77] suggested dividing equation (63) into N classes to classify the range of bubble sizes that may be present within the flow volume, viz.,

$$\frac{\partial n_i}{\partial t} + \nabla \cdot \left(\vec{u}_g n_i\right) = \left(\sum_j R_j\right)_i + \left(R_{ph}\right)_i \tag{64}$$

where $\left(\sum_j R_j\right)_i$ represents the net change in the number density distribution due to coalescence and break-up processes. This interaction term $\left(\sum_j R_j\right)_i$ (= $P_C + P_B$ − D_C − D_B) contains the source rates of P_C, P_B, D_C and D_B, which are respectively, the production rates due to coalescence and break-up and the death rate to coalescence and break-up of bubbles formulated as:

$$P_C = \frac{1}{2}\sum_{k=1}^{N}\sum_{l=1}^{N}\chi_{i,kl}n_i n_j \quad \chi_{i,kl} = \chi_{kl} \text{ if } v_k + v_l = v_i \quad \text{else} \quad \chi_{i,kl} = 0 \text{ if } v_k + v_l \neq v_i$$

$$P_B = \sum_{j=i+1}^{N}\Omega(v_j : v_i)n_j \ ; \ D_C = \sum_{j=1}^{N}\chi_{ij}n_i n_j \ ; \ D_B = \Omega_i n_i \tag{65}$$

To account for the range of bubble classes n_i represented in the vapor phase, the continuity equation of vapor phase is altered to accommodate the additional source terms S_i due to break-up and coalescence mechanisms according to:

$$\frac{\partial \rho_g \alpha_g f_i}{\partial t} + \nabla \cdot (\rho_g \alpha_g \vec{u}_g f_i) = S_i - f_i \Gamma_{lg} \tag{66}$$

The term $f_i \Gamma_{lg}$ represents the mass transfer due to condensation redistributed for each of the discrete bubble classes. The gas void fraction along with the scalar fraction f_i are related to the number density of the discrete bubble ith class n_i (similarly to the jth class n_j) as $\alpha_g f_i = n_i v_i$. The size distribution of the dispersed phase is therefore defined by the scalar f_i.

The term $(R_{ph})_i$ in equation (64) comprises of the essential formulation of the source/sink rates for the phase change processes associated with subcooled boiling flow. At the heated surface, bubbles form at activated cavities known as active nucleation sites. The bubble nucleation rate from these sites can be expressed as:

$$\phi_{WN} = \frac{N_a f \xi_H}{A_C} \tag{67}$$

where ξ_H and A_C are the heated perimeter and the cross-sectional area of the boiling channel respectively. Since the bubble nucleation process only occurs at the heated surface, this heated wall nucleation rate is not included in $\left(R_{ph}\right)_i$ but rather specified as a boundary condition to the continuity equation of vapor phase apportioned to the discrete bubble class n_i based on the bubble departure criteria on the heated surface. The bubble sink rate due to condensation in a control volume for each bubble class can be determined from:

$$\phi_{COND} = -\frac{n_i}{V_B} A_B \frac{dR_B}{dt} \qquad (68)$$

The following holds for the bubble condensation velocity [71]:

$$\frac{dR_B}{dt} = \frac{h\left(T_{sat} - T_l\right)}{\rho_g h_{fg}} \qquad (69)$$

Substituting equation (69) into (68) and given that the bubble surface area A_B and volume V_B based on the Sauter bubble diameter are respectively πD_s^2 and $\pi D_s^3/6$, equation (67) can be rearranged as:

$$\left(R_{ph}\right)_i = \phi_{COND} = -\frac{1}{\rho_g \alpha_g}\left(\frac{6\alpha_g}{D_s}\right)\left[\frac{h\left(T_{sat} - T_l\right)}{h_{fg}}\right] n_i = -\frac{1}{\rho_g \alpha_g}\left[\frac{ha_{if}\left(T_{sat} - T_l\right)}{h_{fg}}\right] n_i \qquad (70)$$

Numerical simulations have been performed in [67] for the experimental setup of Yun et al. [65] and Lee et al. [66] to demonstrate the important effect of equation (70) in a subcooled liquid on the local bubble distribution. The range of wall heat fluxes and associated flow conditions considered for the model are stipulated in figures 13 and 14 respectively. To account for the non-uniform bubble size distribution, bubbles ranging from 0 mm to 9.0 mm were equally divided into 15 classes. This resulted in 15 continuity equations for the vapor phase coupled with a single continuity equation of the liquid phase. Instead of considering 16 different complete phases, it was assumed for the sake of computational efficiency that each bubble class traveled at the same mean algebraic velocity to reduce the computational time. Only a single momentum equation was thus solved for the vapor phase. Prince and Blanch [73] have also indicated additional mechanisms that could be accounted for bubble coalescence such as buoyancy and laminar shear collisions. Indeed collisions caused by buoyancy could not be accounted as it has been assumed that all the bubbles from each class traveled at the same speed. Calculations have also shown that laminar shear collisions were negligible because of the low superficial gas velocities in the range of conditions considered for the subcooled boiling flow investigations.

The distributions of the net rate $\sum_{j=1}^{4} R_j$ (= $P_C + P_B - D_C - D_B$) and condensation rate R_{ph} of Equation (54) are presented in Figure 12. Near the heated wall, the net rate of $\sum_{j=1}^{4} R_j$ being higher than the condensation rate indicated a positive source contribution for the increase in bubble size. Nevertheless, further away from the heated surface, with the condensation rate significantly higher than the net rate and the phenomenon dominating between $(r-R_i)/(R_o-R_i) = 0.2$ and $(r-R_i)/(R_o-R_i) = 0.5$, the model predicted the gradual fragmentation of bubbles as they migrated towards the unheated wall. Figure 14 details the contributions of the production rates of P_C and P_B and death rates of D_C and D_B as well as the condensation rate R_{ph}. The dominant rate near the heated wall was the production rate due to coalescence. An important contribution through the death rate due to coalescence was also noted. The significance of the former indicated a source due to the merging of smaller bubbles forming bigger bubbles. Nevertheless, the latter indicated a source contribution due to the breakage of larger bubbles. The dominant coalescence rates found near the heated wall corresponded to the proper modeling of the physical bubble mechanisms that have been observed in experiments (see Figure 9). The condensation rate gradually diminishing away from the heated wall correctly modeled the condensation phenomenon as also observed through the experimental photographs and confirmed through Gopinath's [71] observations (see Figure 10).

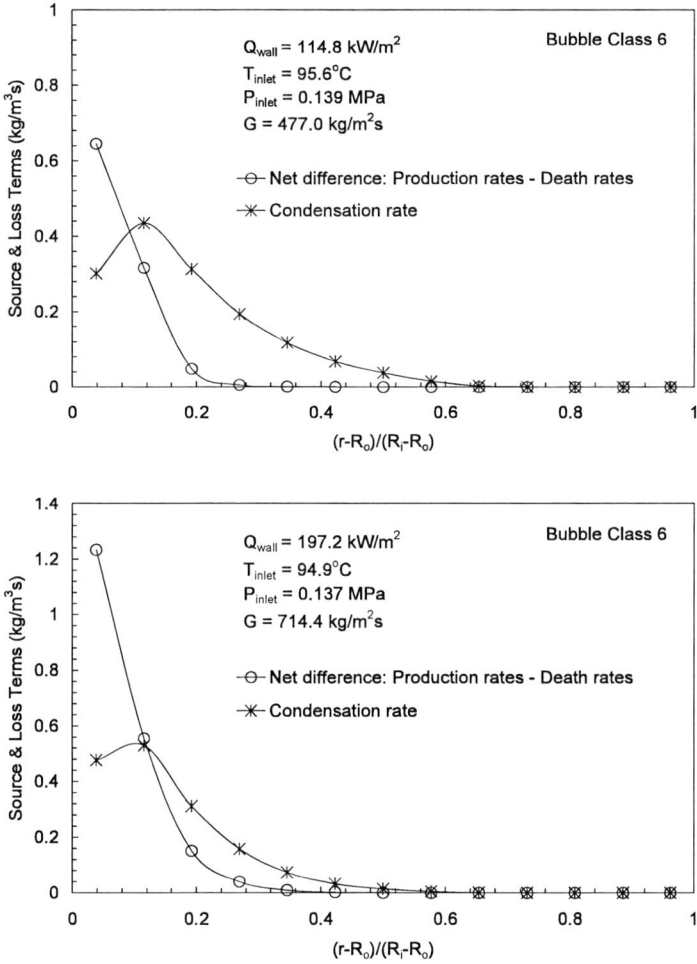

Figure 13. Radial distributions of the net rate $\sum_{j=1}^{4} R_j$ and condensation rate R_{ph} of population balance equation for bubble class 6.

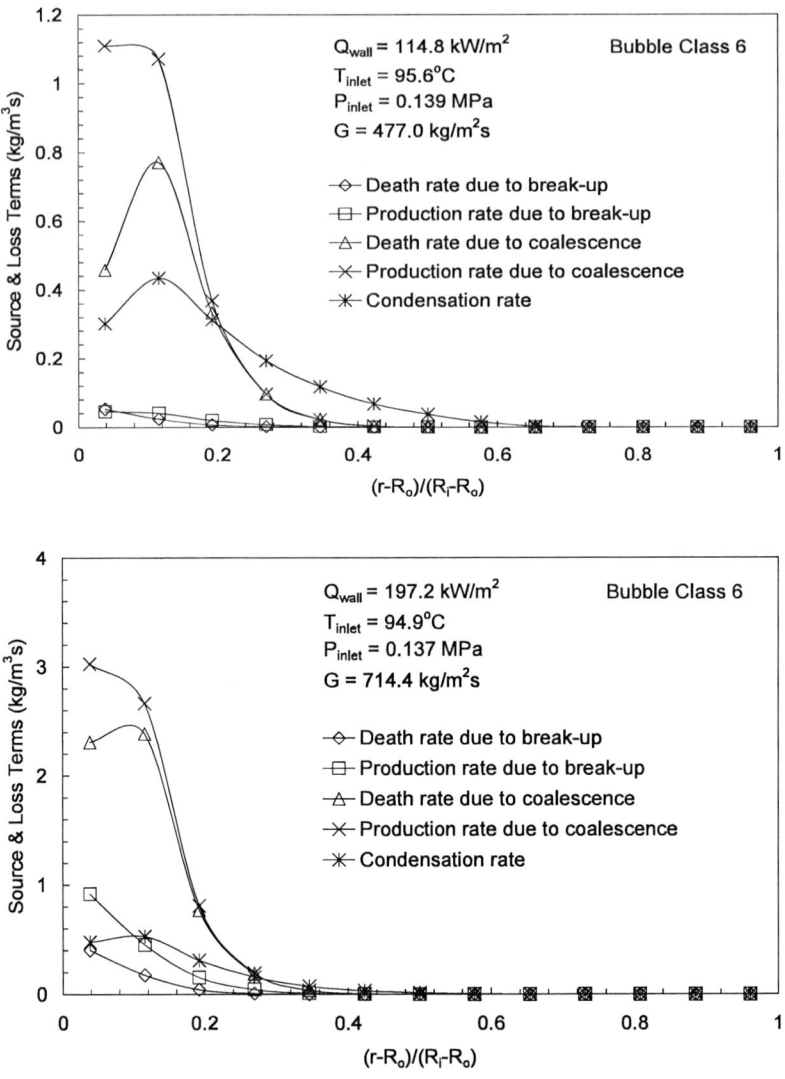

Figure 14. Radial contributions of the production rates of P_C and P_B and death rates of D_C and D_B as well as the condensation rate R_{ph} of population balance equation for bubble class 6.

Chapter 6

IMPROVEMENTS TO WALL HEAT FLUX PARTITIONING MODEL

For vertical upward forced convective boiling flow as described by the experiment shown in Figure 9, it has been observed that bubbles have a tendency to slide before lifting off into the bulk subcooled liquid. In the case where the bubbles may merge with other nucleating bubbles whilst sliding, a smaller number of bubbles will be lifted off from the heated surface area than the actual number of active nucleation sites. As already indicated in Basu et al. [39,40] and reviewed in Warrier and Dhir [41] and Sateesh et al. [77], the transient conduction in such a case due to sliding bubbles becomes the dominant mode of heat transfer. It is therefore imperative that the wall heat partition model incorporates the heat transfer component due to these sliding bubbles.

In order to remove the application uncertainty of correlations empirically determined for the departing bubble diameter, a force balance model has been developed in Yeoh & Tu [78]. It has been demonstrated that the bubble growth and its maximum detached bubble size contributed significantly towards the void growth and heat transfer within the bulk liquid.

As observed through equations (43)-(47), numerous studies have attempted to tie the bubble departure frequency to some other parameter, namely the departing bubble diameter. Although the relationship between the bubble departure frequency and departing bubble diameter offers an attractive means of determining the frequency, application of these relationships, predominantly correlated from pool boiling data, remains contentious for forced convective boiling flow. For example, Cole's bubble frequency relationship (equation (43)) is frequently employed in many investigations of subcooled boiling flows. In order to eliminate the uncertainty of evaluating the bubble frequency, it is proposed that the fundamental theory based on the description of an ebullition cycle in nucleate boiling, which are: (1) waiting period t_w (transient conduction of heat to liquid); (2) growth period t_g: (a) bubble growth rate, (b) evaporation process, (c) agitation of liquid around the bubble and (d) termination of bubble whether by departure or collapse, is employed instead, *viz.*,

$$f = \frac{1}{t_g + t_w} \tag{71}$$

Part of the improvements made to the wall heat partition model to be further described in detail below incorporates the sliding period as well as the fundamental bubble frequency theory of equation (71). When combined with the force balance model, the departing bubble diameter is determined mechanistically. In general, for a given heat flux, a high frequency precedes small bubble diameter, while a low frequency usually results in large bubble diameter

WALL HEAT FLUX PARTITIONING MODEL INCORPORATING BUBBLE SLIDING

Enhancement in heat transfer during forced convective boiling can be attributed by the presence of both sliding and stationary bubbles. There are essentially two mechanisms: (i) The latent heat transfer due to microlayer evaporation and (ii) Transient conduction as the disrupted thermal boundary layer reforms during the waiting period (i.e. incipience of the next bubble at the same nucleation site).

Transient conduction occurs in regions at the point of inception and in regions being swept by sliding bubbles. For a stationary bubble, the heat flux is Q_{tc} given by

$$Q_{tc} = 2\sqrt{\frac{k_l \rho_l C_{pl}}{\pi t_w}}(T_s - T_l) R_f N_a \left(K \frac{\pi D_d^2}{4} \right) t_w f + \\ 2\sqrt{\frac{k_l \rho_l C_{pl}}{\pi t_w}}(T_s - T_l) R_f N_a \left(\frac{\pi D_d^2}{4} \right)(1 - t_w f) \tag{72}$$

where D_d is the bubble departure diameter, T_s is the temperature of the heater surface and T_l is the temperature of the liquid. Equation (72) indicates that some fraction of the nucleation sites will undergo transient conduction while the remaining will be in the growth period. For a sliding bubble, the heat flux Q_{tcsl} due to transient conduction that takes place during the sliding phase and the area occupied by the sliding bubble at any instant of time is given by

$$Q_{tcsl} = 2\sqrt{\frac{k_l \rho_l C_{pl}}{\pi t_w}} (T_s - T_l) R_f N_a l_s K D t_w f + \\ 2\sqrt{\frac{k_l \rho_l C_{pl}}{\pi t_w}} (T_s - T_l) R_f N_a f t_{sl} \left(\frac{\pi D^2}{4}\right) (1 - t_w f) \quad (73)$$

where the average bubble diameter D is given by $D = (D_d + D_l)/2$ and D_l is the bubble lift-off diameter. Strictly speaking, the Kurul & Podowski's model is only applicable for subcooled boiling flows where bubbles are immediately released into the bulk subcooled liquid hence the absence of the bubble sliding phenomenon. This may be possibly true for pool boiling flows in a horizontal orientation.

The reduction factor R_f appearing in equations (72) and (73) depicts the ratio of the actual number of bubbles lifting off per unit area of the heater surface to the number of active nucleation sites per unit area, viz., $R_f = 1/(l_s/s)$ where l_s is the sliding distance and s is the spacing between nucleation sites. In the present study, it shall be assumed that the nucleation sites are distributed in a square grid and that the bubbles slide only in the direction of the fluid flow [36]. The spacing between nucleation sites can thus be approximated as $s = 1/\sqrt{N_a}$. The factor R_f is obtained alongside with the sliding distance evaluated from the force balance model (to be described below). The significance of this factor provides the information whereby the bubble departing from its site of origin merges with other nucleating bubbles at adjacent sites. It is noted that for the case where the sliding distance l_s is less than the spacing s, $R_f = 1$.

Forced convection will always prevail at all times in areas of the heater surface that are not influenced by the stationary and sliding bubbles. The fraction of the heater area for stationary and sliding bubbles is given by

$$1 - A_q = 1 - R_f \left[N_a \left(K \frac{\pi D_d^2}{4} \right) t_w f + N_a \left(\frac{\pi D_d^2}{4} \right) (1 - t_w f) \right.$$
$$\left. + N_a l_s K D t_w f + N_a f t_{sl} \left(\frac{\pi D^2}{4} \right) (1 - t_w f) \right] \quad (74)$$

FORCE BALANCE MODEL FOR BUBBLE DEPARTURE AND BUBBLE LIFT-OFF

The development of the force balance model concentrates on the various forces that influence the growth of a bubble during flow conditions in the directions parallel and normal to a vertical heating surface. These forces are formulated according to the studies performed by Klausner et al. [79] and Zeng et al. [80]. Figure 15 illustrates the forces acting on the bubble in the *x-direction* and *y-direction*; they are respectively,

$$\Sigma F_x = F_{sx} + F_{dux} + F_{sL} + F_h + F_{cp} \quad (75)$$

and

$$\Sigma F_y = F_{sy} + F_{duy} + F_{qs} + F_b \quad (76)$$

where F_s is the surface tension force, F_{du} is the unsteady drag due to asymmetrical growth of the bubble and the dynamic effect of the unsteady liquid such as the history force and the added mass force, F_{sL} is the shear lift force, F_h is the force due to the hydrodynamic pressure, F_{cp} is the contact pressure force accounting for the bubble being in contact with a solid rather than being surrounded by liquid, F_{qs} is the quasi steady-drag in the flow direction, and F_b is the buoyancy force. In addition, g indicates the gravitational acceleration; α, β and θ_i are the advancing, receding and inclination angles respectively; d_w is the surface/bubble contact diameter; and d is the vapor bubble diameter at the wall.

The forces acting in the *x-direction* can be estimated from:

$$F_{sx} = -d_w \sigma \frac{\pi}{\alpha - \beta} [\cos \beta - \cos \alpha]; \quad F_{dux} = -F_{du} \cos \theta_i;$$

$$F_{sL} = \frac{1}{2}C_L\rho_l\Delta U^2\pi r^2 \;;\; F_h = \frac{9}{4}\rho_l\Delta U^2\frac{\pi d_w^2}{4} \;;\; F_{cp} = \frac{\pi d_w^2}{4}\frac{2\sigma}{r_r}$$

In the *y-direction*, they are:

$$F_{sy} = -d_w\sigma\frac{\pi(\alpha-\beta)}{\pi^2-(\alpha-\beta)^2}[\sin\alpha+\sin\beta] \;;\; F_{duy} = -F_{du}\sin\theta_i \;;$$

$$F_{qs} = 6C_D\mu_l\Delta U\pi r \;;\; F_b = \frac{4}{3}\pi r^3(\rho_l-\rho_g)g$$

From the various forces described along the *x-direction* and *y-direction*, *r* is the bubble radius, ΔU is the relative velocity between the bubble centre of mass and liquid, C_D and C_L are the respective drag and shear lift coefficients and r_r is the curvature radius of the bubble at the reference point on the surface $x = 0$, which is $r_r \sim 5r$ [79].

The drag coefficient C_D and shear lift coefficient C_L appearing in the drag and lift forces are determined according to the relationships proposed by Klausner et al. [79], viz.,

$$C_D = \frac{2}{3} + \left[\left(\frac{12}{Re_b}\right)^n + 0.796^n\right]^{-1/n} \tag{77}$$

$$C_L = 3.877G_s^{1/2}\left[\frac{1}{Re_b^2} + 0.014G_s^2\right]^{1/4} \tag{78}$$

where $n = 0.65$ and $Re_b = \rho_l\Delta Ud/\mu_l$ is the bubble Reynolds number. The dimensionless shear rate G_s is: $(dU/dx)(r/\Delta U)$. The gradient dU/dx can be determined through the universal velocity profile for turbulent flow:

$$\frac{U}{u_\tau} = 2.5\ln(9.8x^+) \tag{79}$$

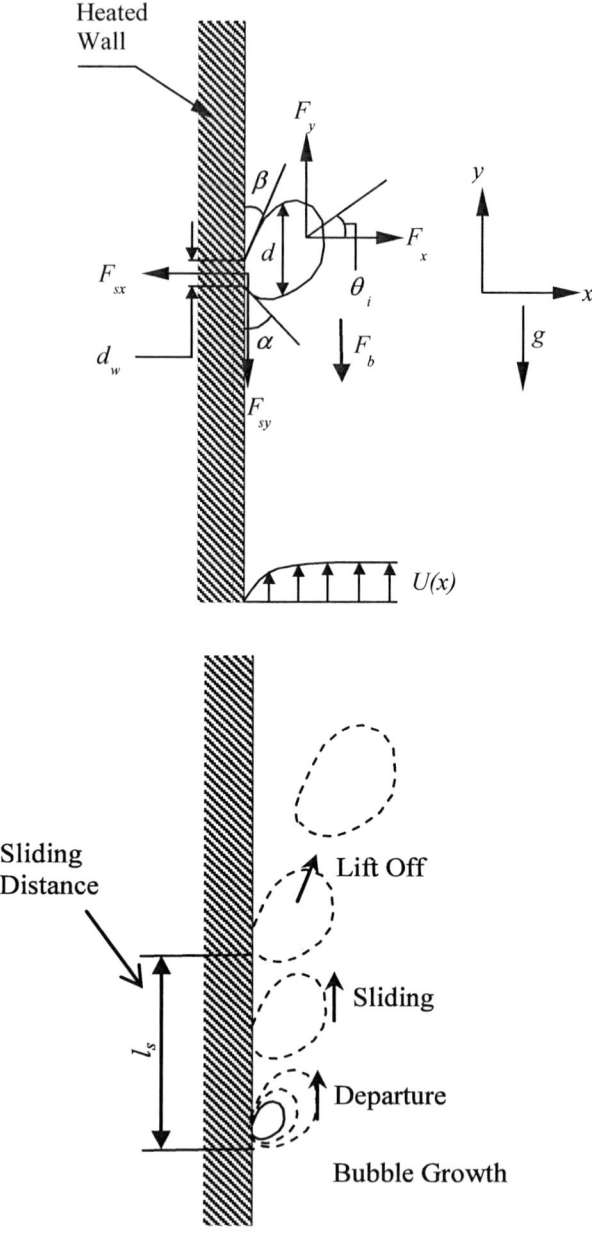

Figure 15. Schematic drawings illustrating the forces acting on a growing vapor bubble (above) and a bubble departing, sliding and lifting off from a vertical heated surface (below).

where u_τ is the friction velocity and $x^+ = \rho_l u_\tau x / \mu_l$ is the non-dimensional normal distance from the heated wall. In equation (79), the velocity profile is assumed to be applicable for the time-averaged velocity distribution in the vicinity of the heated wall. Adjacent velocities, determined through the two-fluid model, are used to obtain the varying local friction velocities through equation (79). These friction velocities are subsequently used to evaluate the gradients dU/dx along the heated wall to determine the shear rate G_s.

The growth force F_{du} is modeled by considering a hemispherical bubble expanding in an inviscid liquid, which is given by Zeng et al. [80] as

$$F_{du} = \rho_l \pi r^2 \left(\frac{3}{2} C_s \dot{r}^2 + r\ddot{r} \right) \tag{80}$$

where (˙) indicates differentiation with respect to time. The constant C_s is taken to be 20/3 according to [80]. In estimating the growth force, additional information on the bubble growth rate is required. As in [80], a diffusion controlled bubble growth solution by Zuber [81] is adopted:

$$r(t) = \frac{2b}{\sqrt{\pi}} Ja \sqrt{\eta t} \; ; \; Ja = \frac{\rho_l C_{pl} \Delta T_w}{\rho_g h_{fg}} \; ; \; \eta = \frac{\lambda_l}{\rho_l C_{pl}} \tag{81}$$

where Ja is the Jakob number, η is the liquid thermal diffusivity and b is an empirical constant that is intended to account for the asphericity of the bubble. For the range of heat fluxes investigated in this investigation, b is taken to be 0.21 based on a similar subcooled boiling study performed by Steiner et al. [38], which was experimentally verified through their in-house measurements with water as the working fluid.

While a vapor bubble remains attached to the heated wall, the sum of the parallel and normal forces must satisfy the following conditions: $\Sigma F_x = 0$ and $\Sigma F_y = 0$. For a sliding bubble case, the former establishes the bubble departure diameter (D_d) while the latter yields the bubble lift-off diameter (D_l). The growth period t_g appearing in equation (72) can be readily evaluated based on the availability of the bubble size at departure from its nucleation site through equation (70). The lift off period t_l can also be similarly calculated based on the bubble lift-off diameter. The difference between the bubble lift-off and bubble growth periods provides the period for the sliding bubble; the sliding distance l_s

can subsequently be determined (see Figure 14). An estimation on this sliding distance can be determined according to the experimental correlation of Maity [82] as $l_s = (2/3)C_v t_{sl}^{3/2}$ where t_{sl} is the sliding time ($t_l - t_g$) and C_v is an acceleration coefficient correlated in terms of the tangential liquid velocity (u_l) adjacent to the heated surface: $C_v = 3.2u_l + 1$. This coefficient reflects the increase in bubble velocity with time after it begins to slide away from a nucleation site.

The bubble waiting time t_w is determined through the occurrence of transient conduction when a bubble slides or lifts off of which the boundary layer gets disrupted and cold liquid comes in contact with the heated wall. Assuming that the heat capacity of the heater wall $\rho_s C_{ps} \delta_s$ is very small, the conduction process can be modeled by considering one-dimensional transient heat conduction into a semi-infinite medium with the liquid at a temperature T_l and the heater surface at a temperature T_s. The wall heat flux can be approximated by

$$Q_w = \frac{k_l (T_s - T_l)}{\delta_l} \tag{82}$$

where δ_l ($=\sqrt{\pi \eta t}$) is the thickness of the thermal boundary layer. If the temperature profile inside this layer is taken to be linear according to Hsu & Graham [80], it can thus be expressed as

$$T_b = T_w - \frac{(T_s - T_l)x}{\delta_l} \tag{83}$$

where x is the normal distance from the wall. Based on the criterion of the incipience of boiling from a bubble site inside the thermal boundary layer, the bubble internal temperature for a nucleus site (cavity) with radius r_c is

$$T_b = T_{sat} + \frac{2\sigma T_{sat}}{C_2 r_c h_{fg} \rho_g} \quad \text{at } x = C_1 r_c \tag{84}$$

where $C_1 = (1 + \cos\theta)/\sin\theta$ and $C_2 = 1/\sin\theta$. The angle θ refers here represents the bubble contact angle as described previously. By substituting equation (84) into equation (83), the waiting time t_w can be obtained as

$$t = t_w = \frac{1}{\pi \eta} \left[\frac{(T_s - T_l) C_1 r_c}{(T_w - T_{sat}) - 2\sigma T_{sat}/C_2 \rho_g h_{fg} r_c} \right]^2 \qquad (85)$$

The cavity radius r_c can be determined by applying Hsu's criteria and tangency condition of equations (83) and (84), viz.,

$$t = \left[\frac{C_1 C_2 \rho_g h_{fg} r_c^2}{2\sigma T_{sat}} \right]^2 \frac{(T_s - T_l)^2}{\pi \eta} = \left[\frac{k_l}{Q_w} \right]^2 \frac{(T_s - T_l)^2}{\pi \eta} \qquad (86)$$

From the above equation,

$$r_c = F \left[\frac{2\sigma T_{sat} k_l}{\rho_g h_{fg} Q_w} \right]^{1/2} \qquad (87)$$

where

$$F = \left(\frac{1}{C_1 C_2} \right)^{1/2} = \left(\frac{\sin^2 \theta}{1 + \cos \theta} \right)^{1/2}$$

According to Basu et al. [82], the factor F indicates the degree of flooding of the available cavity size and the wettability of the surface. If the contact angle $\theta \rightarrow 0$, all the cavities will be flooded. Alternatively, as $\theta \rightarrow 90°$, $F \rightarrow 1$, all the cavities will not be flooded (i.e. they contain traces of gas or vapor).

In reality, the surface/bubble contact diameter d_w evolves from the point of inception until the point of departure or lift-off. Here, a correlation based on the experimental data of Maity [82] as a function of the bubble contact angle θ is employed to determine the ratio of the bubble base diameter d_w to the bubble diameter at departure or lift-off, which is given as $C = 1 - \exp(-2\theta^{0.6})$. Experimental observations by Klausner et al. [79] and Bibeau and Salcudean [84] have indicated that the advancing angle α and receding angle β varied quite substantially during the sliding phase. Considering the complexity of the bubble departure and bubble lift-off, and the difficulty in obtaining the measurements, the advancing and receding angles can be reasonably evaluated through the bubble contact angle θ as $\alpha = \theta + \theta'$ and $\beta = \theta - \theta'$. Klausner et al. [79] have employed an angle θ' of 4.5° in their theoretical analysis while Bibeau and Salcudean [84] have reported a value of 2.5°. According to Winterton [85], the angle θ' has nonetheless been postulated to be as high as 10°. In the present study, an angle θ' of 5° is adopted. For the inclination angle θ_i, a value of 10° that gave the best fit to the data by Klausner et al. [79] is employed for the current force balance model.

MODEL PREDICTIONS

The improved wall heat flux partitioning model coupled with the force balance model were assessed by comparing the model predictions against subcooled flow boiling of local radial measurements by Yun et al. [65] and Lee et al. [66] and axial measurements by Zeitoun and Shoukri [61]. Based on their experimental observations, the bubble contact angles have been taken to be at 35° for the local cases and 45° for the axial cases respectively. Experimental conditions for the local and axial data that have been used for comparison with the simulated results are presented in table 2. Note that the experimental conditions for the local casse chosen herein are identical to the investigations while applying the empirical correlations in Figure 7.

Table 2. Experimental conditions for local (L1, L2, L3) and axial (A1, A2, A3) cases

	P_{inlet} (MPa)	Q_w (kW/m²)	G (kg/m² s)	T_{inlet} (°C)	$T_{sub,inlet}$ (°C)
L1	0.143	152.9	474.0	96.9	13.4
L2	0.137	197.2	714.4	94.9	13.8
L2	0.143	251.5	1059.2	92.1	17.9
A1	0.137	286.7	156.2	91.9	14.9
A2	0.150	508.0	264.3	94.6	16.6
A3	0.150	705.0	411.7	88.9	22.5

The conservation equations for mass, momentum and energy of each phase were discretised using the finite volume technique. A total number 15 bubble classes were prescribed for the dispersed phases, which are illustrated in table 3 for the local and axial cases. This representd an additional set of 15 transport equations of which they were progressively solved and coupled with the flow equations during the simulations.

Table 3. Evaluated diameters of each discrete bubble classes for the local and axial cases

Bubble class	d_{local} (mm)	d_{axial} (mm)
1	0.45	0.29
2	0.94	0.61
3	1.47	0.95
4	2.02	1.31
5	2.58	1.67
6	3.14	2.03
7	3.71	2.40
8	4.27	2.76
9	4.83	3.13
10	5.40	3.49
11	5.96	3.86
12	6.53	4.23
13	7.10	4.59
14	7.66	4.96
15	8.23	5.32

Numerical calculations were performed for one-quarter of the cross-section, utilizing the symmetry property of the annulus geometry for the local as well as the axial cases. A body-fitted conformal system was employed to generate the three-dimensional mesh within the annular channel resulting in a total of 13 (radial) × 30 (height) × 3 (circumference) control volumes for the local case while a total of 8 (radial) × 20 (height) × 3 (circumference) control volumes were used for the axial case. Since wall function was used in the present study to bridge the wall and the fully turbulent region away from heater surface, the normal distance between the wall and the first node in the bulk liquid should be such that the corresponding x^+ was greater than 30. Grid independence was examined. In the mean parameters considered, further grid refinement did not reveal significant changes to the two-phase flow parameters.

For the local cases of L1, L2 and L3, the measured and predicted radial profiles of the Sauter mean bubble diameter (D_s), vapor void fraction (α_g) and interfacial area concentration (a_{if}) located at the measuring plane 1.61 m downstream of the beginning of the heated section are shown in figures 16-18. Note again that in all the figures, the dimensionless parameter $(r-R_i)/(R_o-R_i) = 1$ indicates the inner surface of the unheated flow channel wall while $(r-R_i)/(R_o-R_i) = 0$ indicates the surface of the heating rod in the channel.

Experimental photographs (see Figure 9) clearly showed large bubble sizes being present away from the heated wall but not at the heated wall, which this trend has been correctly modeled and predicted by the population balance model.

The presence of larger bubbles away from the heated wall confirmed the occurrence of bubble coalescence. In all the three cases, good agreement was achieved between the predicted and measured profiles with the maximum predicted Sauter bubble diameters for L1, L2 and L3 were obtained at about 4.9 mm, 4.7 mm and 3.9 mm while the experimental maximum bubble sizes were measured at about 4.7 mm, 4.9 mm and 4.1 mm. Further away from the heated wall, the reduction of the bubble sizes indicated the bubbles condensing due to the subcooled bulk liquid. It should be noted that the local determination of the Sauter bubble diameter by the population balance model required the appropriate determination of the bubble lift-off diameters through the force balance model. At the measuring location, the bubble departure diameters were predicted to be about 0.56 mm, 0.58 mm and 0.59 mm while the bubble lift-off diameters were approximately attained as 1.4 mm 1.3 mm and 1.2 mm respectively. From these bubble sizes, the ratio of the bubble lift-off diameter to the bubble departure diameter achieved an approximate value of two, which also showed good agreement against experimental observations of Basu et al. [39]. For the bubble frequency, the predicted growth and waiting times for L1 were found to yield about 5.3 ms and 0.74 ms at wall superheat and subcooling temperatures of 18.1°C and 4.1°C, L2, about 5.1 ms and 0.63 ms at wall superheat and subcooling temperatures of 18.5°C and 4.5°C, and L3, about 4.9 ms and 0.67 ms at wall superheat and subcooling temperatures of 19.6°C and 8.8°C. These predicted growth and waiting times were of a similar order according to the values correlated by Basu et al. [36] though they have exerted that the application of their correlated relationships should be used with care since they have been formulated through a limited database. At the point of bubble lifting off into the bulk liquid, the shear lift force was seen as the dominant force in overcoming the surface tension forces acting on the vapor bubble. The calculated shear lift and surface tension forces were respectively:

L1: $F_{sx} \sim 5.06 \times 10^{-5}$ N and $F_{sL} \sim 1.78 \times 10^{-4}$ N;
L2: $F_{sx} \sim 5.06 \times 10^{-5}$ N and $F_{sL} \sim 1.76 \times 10^{-4}$ N; and
L3: $F_{sx} \sim 5.06 \times 10^{-5}$ N and $F_{sL} \sim 1.73 \times 10^{-4}$ N.

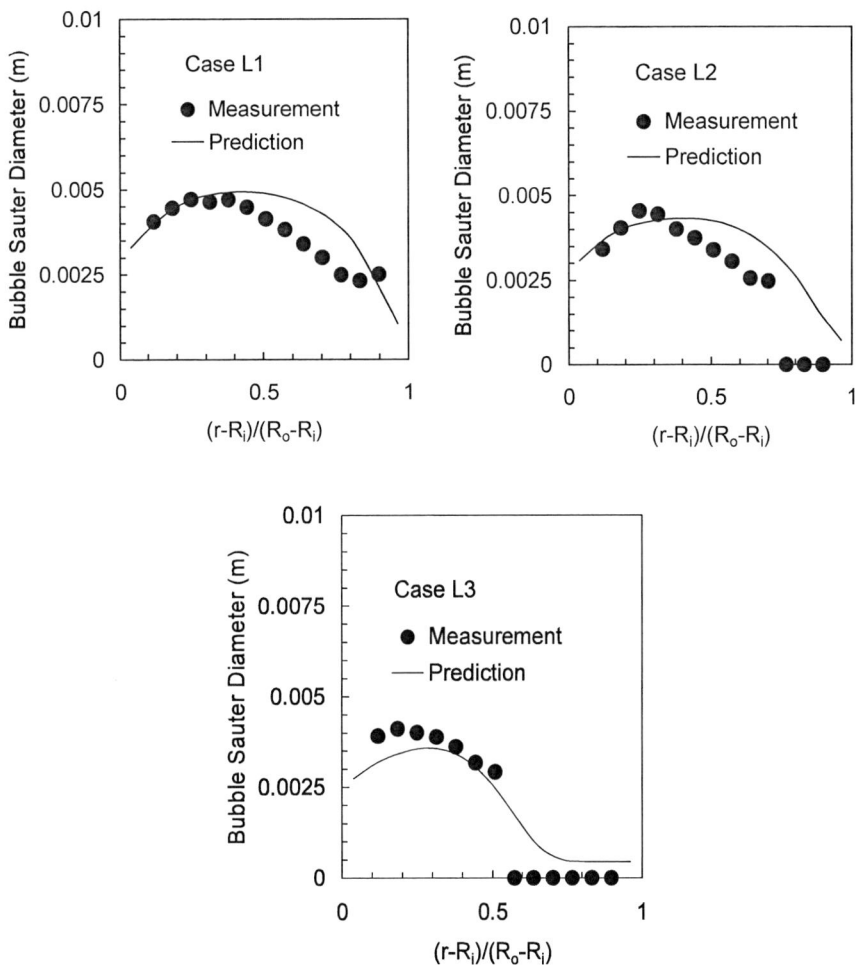

Figure 16. Local Sauter bubble diameter profiles at the measuring plane.

Peak local void fractions observed near the heated surface in Figure 17 represented a typical chracteristic commonly found in subcooled boiling flow. This high local void fraction found there was explicitly due to the large number of bubbles generated from the active nucleation sites on the heated surface. Here, large amount of bubbles were generated from these active nucleation sites when the temperature on the heated surface exceeded the saturation temperature. A proper prediction of the active nucleation site densities was therefore crucial. As these bubbles reached a critical size, they detached from the heated surface and migrated laterally toward the subcooled liquid core under the competing process of bubble coalescence and condensation as aforementioned.

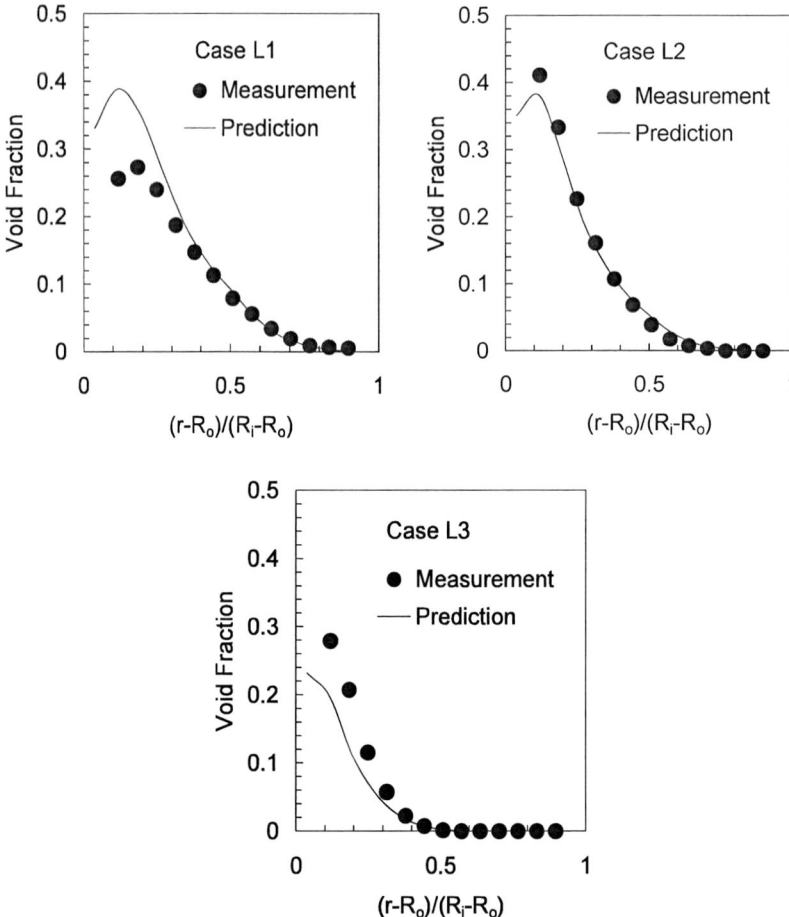

Figure 17. Local void fraction profiles at the measuring plane.

Figure 18 describes the local interfacial area concentration radial distribution. Good agreement was achieved between the measured and predicted interfacial area concentration for all the cases except for case L1 where the interfacial area concentration was significantly over predicted near the heated wall. This could be inferred by the high void fraction predicted near the heated wall as seen in Figure 17. The trend of the interfacial area concentration distributions for all the three cases were correctly modeled and predicted by the two-fluid and population balance models.

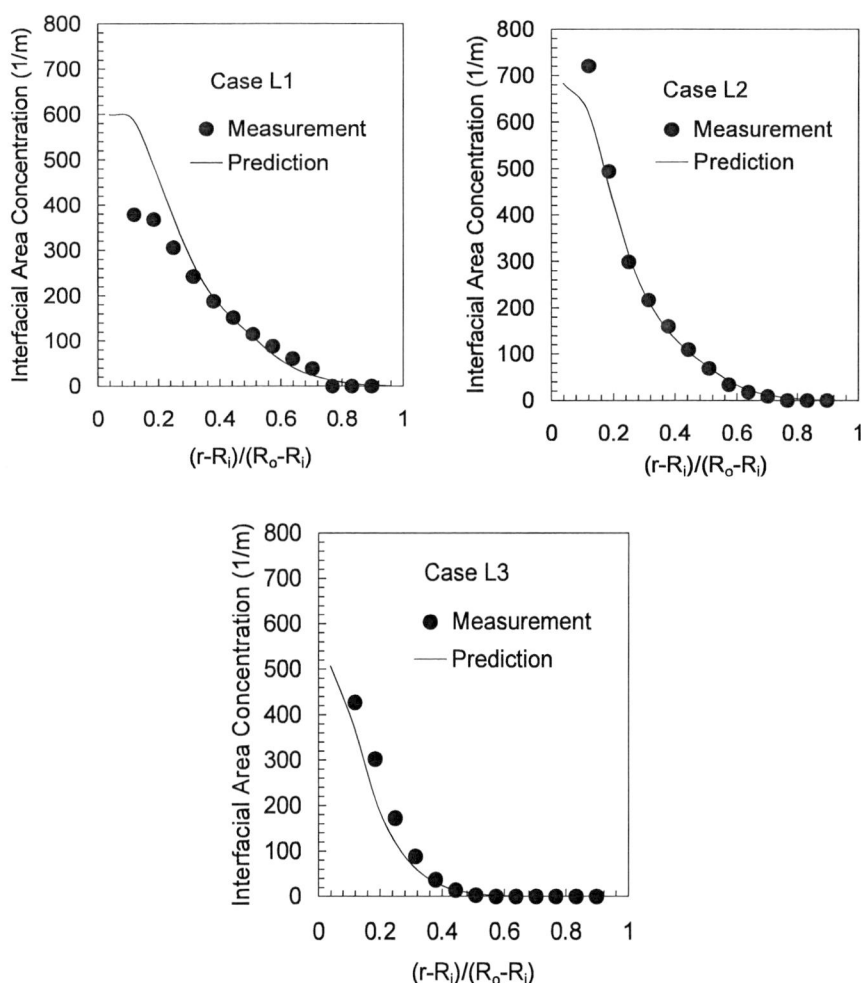

Figure 18. Local interfacial area concentration profiles at the measuring plane.

The radial profiles of the axial component of the local vapor velocity are shown in Figure 19 while Figure 20 presents the radial profiles of the local liquid velocity for experimental conditions L1, L2 and L3.

The vapor velocity being greater than the liquid velocity was due to buoyancy force caused by density difference. As observed in the experiment, the vapor velocity was higher at the centre than the velocities near the heating rod. This was probably due to the buoyancy effect being enhanced by the migration of the large bubbles there, which was again confirmed by high-speed photography in Lee et al. [66]. Vapor velocity predicted by the model showed that higher velocity values approaching the heated boundary. This was mainly due to the assumption where each bubble class has been considered to be traveling at the same mean algebraic velocity due to a single momentum equation being solved for the vapor phase. The philosophy behind adopting this approach was to hasten the computational time and to reduce the computational resources. The discrepancies between the predicted and measured velocities found near the heated wall clearly demonstrated the inadequacy of the adopted approach.

Within the channel space, different size bubbles can be expected to travel with different speeds. As an initial step towards resolving the problem, an algebraic slip model could be proposed to account for bubble separation. The terminal velocities for each of the bubbles can be considered through applying an algebraic relationship suggested by Clift et al. [83], which could be then used to evaluate the individual bubble slip velocities. Alternatively, the consideration of additional momentum equations to cater for each of the 15 bubble classes would increase the computational resources tremendously and deem impractical. Ongoing investigations are currently undertaken to test a pertinent choice of two or three dominant groups of bubbles transformed into the Eulerian phases to sufficiently accommodate the hydrodynamics of wide bubble size distributed bubbly flows such as the recent development of the inhomogeneous model of the population balance approach based on the multiple size group (MUSIG) in ANSYS-CFX11. Nevertheless, good agreement was achieved for the liquid velocities between the predictions and experimental values at the measuring plane in the liquid phase. These velocities showed a closer resemblance to the measurements than the predicted profiles of the vapor velocity.

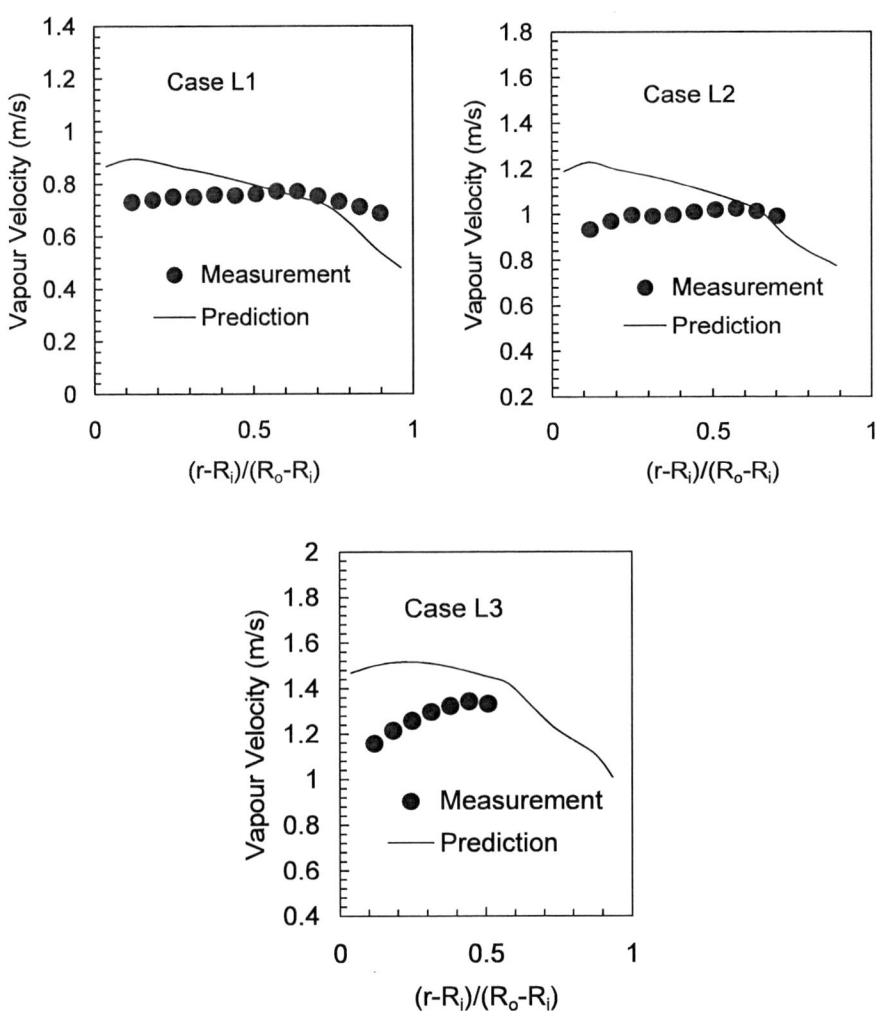

Figure 19. Local vapor velocity profiles at the measuring plane.

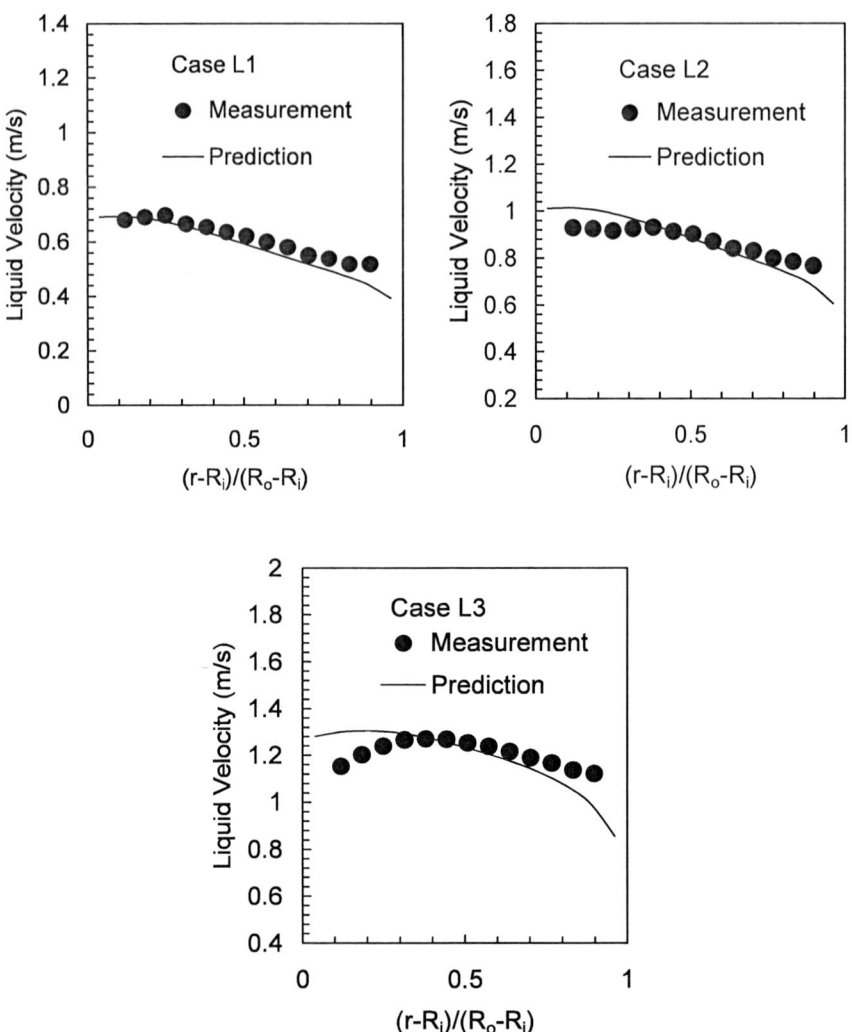

Figure 20. Local liquid velocity profiles at the measuring plane.

For the axial Cases A1, A2 and A3, the measured and predicted profiles of Sauter bubble diameter normalized by the length scale $\sqrt{\sigma/g(\rho_l - \rho_g)}$, interfacial area concentration and the void fraction along the heated section are presented in figures 21-23.

The increasing bubble sizes along the axial length clearly showed the bubble coalescence intensifying as the bubbles traveled downstream. Predicted Sauter bubble diameter profiles were seen to agree well with the experimental data as shown in Figure 20. At a point before the channel exit, the predicted growth and waiting times for A1 were ascertained to be about 4.6 ms and 0.5 ms at wall superheat and subcooling temperatures of 18.9°C and 4.3°C; A2, about 4.8 ms and 0.23 ms at wall superheat and subcooling temperatures of 22.8°C and 2.3°C; and A3, about 4.8 ms and 0.24 ms at a wall superheat and subcooling temperatures of 23.0°C and 7.0°C. As expected, the waiting times especially for A2 and A3 were much lower than A1 due to the higher wall heat fluxes.

Figure 22 describes the axial interfacial area concentration distribution while Figure 23 demonstrates the axial void fraction evolution along the heated section. Owing to the good agreement achieved between the predictions and measurements of the void fraction, the interfacial area concentration were also found to be in good agreement with the measured values, for the given axial bubble size evolution. The predicted void fraction profiles clearly represented the subcooled boiling characteristic as depicted in Figure 1: low void fraction region followed by a second region, in which the void fraction increased significantly. The predicted void fraction profiles also correctly established a plateau at the initial stages of boiling. The sharp increase of the void fraction profiles approaching the channel exit was due to the increase of bubble sizes decreasing the interfacial area concentration thereby causing a reduction in the condensation rate per unit volume of the channel. This was also associated with the reduction in the relative importance of condensation at the bubble interface due to the decrease of liquid subcooling, which encouraged more bubbles to merge.

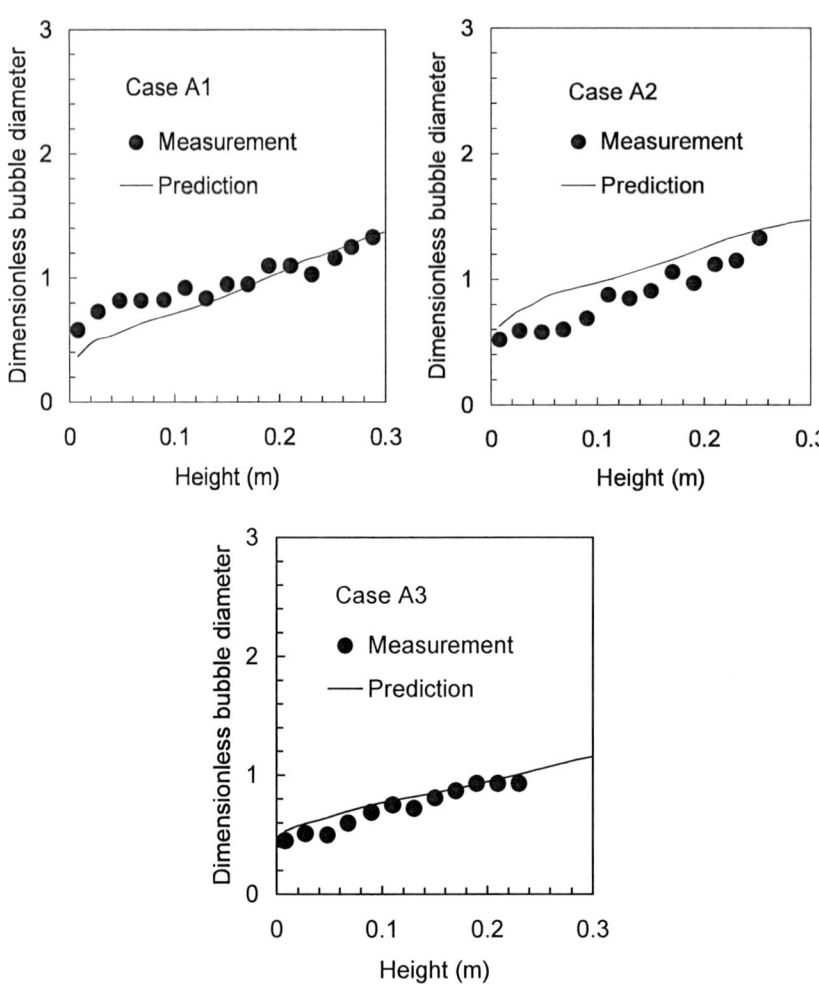

Figure 21. Axial dimensionless Sauter bubble diameter void fraction profiles.

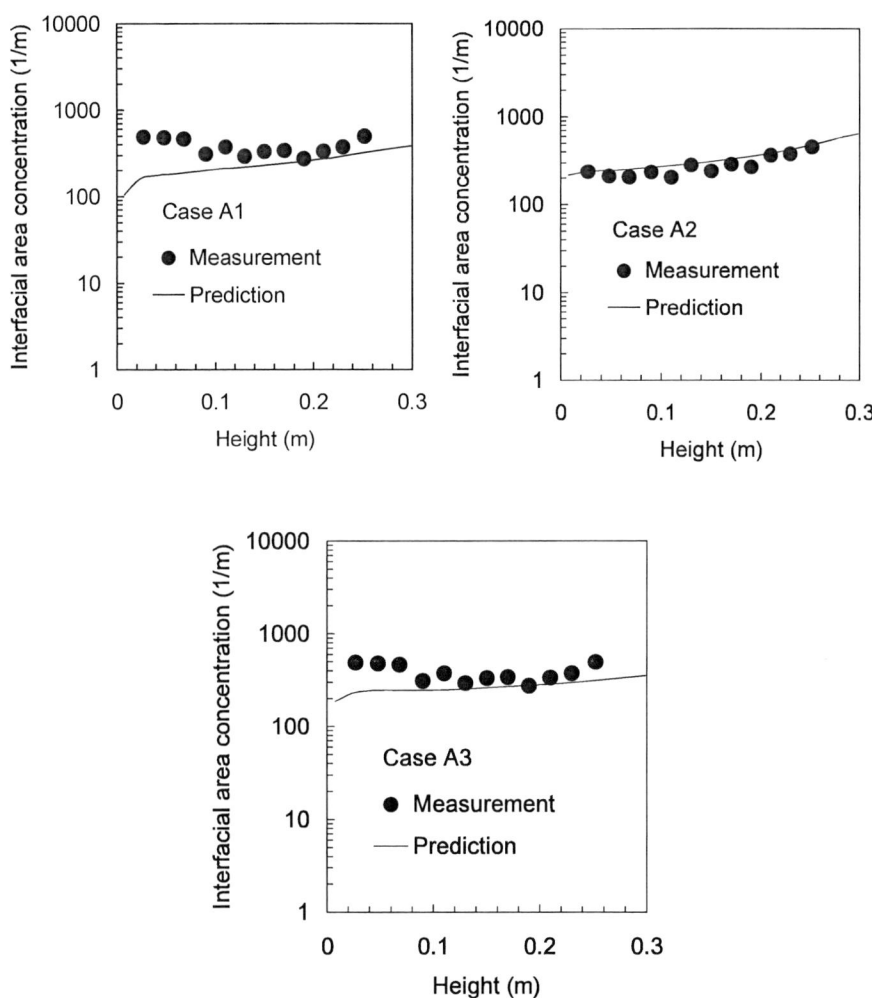

Figure 22. Axial interfacial area concentration profiles.

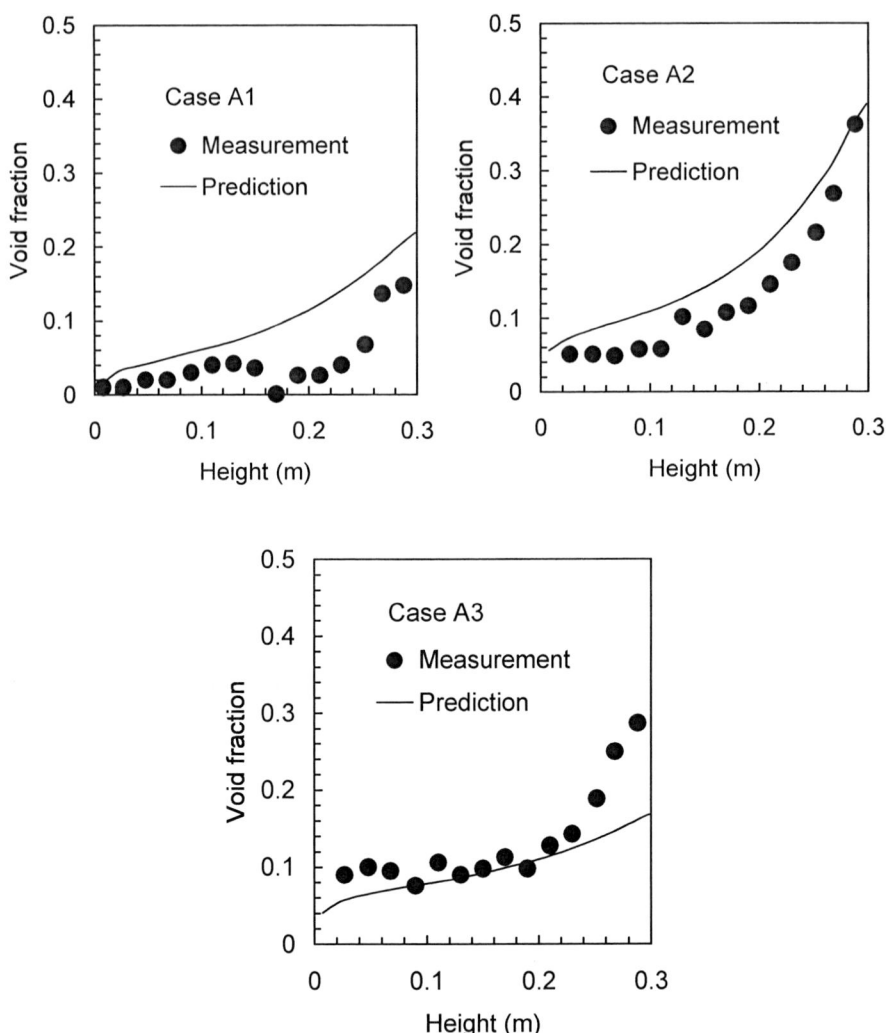

Figure 23. Axial void fraction profiles.

Chapter 7

SUMMARY

A review of both empirical correlations and mechanistic models to handle subcooled boiling flow has been presented. Empirical correlations which are usually based on curve fitting experimental datasets are unable to account for the physical mechanisms that are unherently present within the complex boiling process. Significant differences between the predictions and experimental data can thus occur when the conditions for which they were developed are not duplicated. This represents the major drawback of the application of such models.

In this article, a two-fluid model coupled with population balance approach is presented for handling low-pressure subcooled boiling flow. The increase in complexity of modeling such flow derives from the additional consideration of the gas or liquid undergoing a phase transformation. Formulation of an improved wall heat flux partitioning model that account for sliding bubbles alongside with the fundamental consideration of bubble frequency theory during low-pressure subcooled flow boiling is also presented. This model when coupled with a force balance model to predict forces acting on a vapor bubble growing under subcooled boiling flow demonstrates the capability to accommodate more complex analyses of bubble growth, bubble departure and bubble lift-off over a wide range of wall heat fluxes and flow conditions. When assessed against local and axial measurements, this present model reveals remarkable agreement for the local and axial profiles of the void fraction, Sauter bubble diameter and interfacial area concentration for different experimental conditions. For the local case, good agreement is also achieved by the present model with regards to the radial distributions of the liquid velocity profiles. Significant weakness is however found for the vapor velocity distribution, which requires additional modeling. For the axial case, a plateau occurring at the initial boiling stages, typical of subcooled flow boiling at low pressures, followed by the significant rise of the void fraction downstream of the annular channel is predicted by the present model.

REFERENCES

[1] Bergles, A. E., & Rohsenow, W. M. (1964). The determination of forced convection, surface boiling heat transfer, *ASME J. Heat Transfer*, *86*, 365-372.

[2] Mikic, B. B., & Rohsenow, W. M. (1969). A new correlation of pool boiling data including the effect of heating surface characteristics, *ASME J. Heat Transfer*, *91*, 245-256.

[3] Chen, J. C. (1966). A correlation for boiling heat transfer to saturated fluids in convective flows, *Ind. Eng. Chem. Process Des. Dev.*, *5*, 322-329.

[4] Kandlikar, S. G. (1998). Heat transfer characteristics in partial boiling, fully developed boiling, and significant void flow regions of subcooled flow boiling, *ASME J. Heat Transfer*, *120*, 395-401.

[5] McAdams, W. H., Kennel, W. E., Minden, C. S., Carl, R., Picornell, P. M., & Dew, J. E. (1949). Heat transfer at high rates to water with surface boiling, *Ind. Eng. Chem.*, *41*, 1945-1953.

[6] Thom, J. R. S., Walker, W. M., Fallon, T. A., & Reising, G. F. S. (1965). Boiling in subcooled water during flow up heated tubes or annuli, presented at the Symposium on Boiling Heat Transfer in Steam Generating Units and Heat Exchangers, Institute of Mechanical Engineers, London.

[7] Griffith, P., Clark, J. A., & Rohsenow, W. M. (1958). Void volumes in subcooled boiling, *ASME Paper 58-HT-19.*, U.S. National Heat Transfer Conference, Chicago.

[8] Zuber, N., & Findlay, J. (1965). Average volumetric concentration in two-phase flow systems, *ASME J. Heat Transfer*, *87*, 453-462.

[9] Lahey, R. T. Jr. (1978). A mechanistic subcooled boiling model, Proceedings of the 6th International Heat Transfer Conference, Toronto, Canada, Hemisphere Publishing Corporation, Washington, *1*, 293-297.

[10] Chatoorgoon, V., Dimmick, G. R., Carver, M. B., Selander, W. N., & Shoukri, M. (1992). Application of generation and condensation models to predict subcooled boiling void at low pressure, *Nuc. Technol.*, *98*, 366-378.

[11] Zeitoun, O. (1994). Subcooled flow boiling and condensation, Ph.D. Thesis, McMaster University, Hamilton, Ontario, Canada.

[12] Krishna, R., Urseanu, M. I., van Baten, J. M., & Ellenberger J. (1999). Influence of scale on the hydrodynamics of bubble columns operating in the churn-turbulent regime: experiments vs. Eulerian simulations, *Chem. Eng. Sci.*, *54*, 4903-4911.

[13] Shimizu, K., Takada, S., Minekawa, K., & Kawase, Y. (2000). Phenomenological model for bubble column reactors: prediction of gas hold-up and volumetric mass transfer coefficients, *Chem. Eng. J.*, *78*, 21-28.

[14] Pohorecki, R., Moniuk, W., Bielski, P., & Zdrojkwoski, A. (2001). Modeling of the coalescence/redispersion processes in bubble columns, *Chem. Eng. Sci.*, *56*, 6157-6164.

[15] Olmos, E., Gentric, C., Vial Ch., Wild, G., & Midoux, N. (2001). Numerical simulation of multiphase flow in bubble column reactors. Influence of bubble coalescence and break-up, *Chem. Eng. Sci.*, *56*, 6359-6365.

[16] Ramkrishna, D., & Mahoney, A. W. (2002). Population balance modeling. Promise for the future, *Chem. Eng. Sci.*, *57*, 595-606.

[17] Ranz, W. E., & Marshall, W. R. (1952). Evaporation from droplets: parts I and II, *Chem. Eng. Prog.*, *48*, 141-148.

[18] Ishii, M., & Zuber, N. (1979). Drag coefficient and relative velocity in bubbly, droplet or particulate flows, *AIChE J.*, *25*, 843-855.

[19] Drew, D. A., & Lahey, R. T. Jr. (1979). Application of general constitutive principles to the derivation of multi-dimensional two-phase flow equation, *Int. J. Multiphase Flow*, *5*, 243-264.

[20] Lopez de Bertodano, M. (1992). Tubulent bubbly two-phase flow in a triangular duct, Ph.D. Thesis, Rensselaer Polytechnic Institute, Troy, New York.

[21] Antal, S. P, Lahey, R. T. Jr., & Flaherty, J. E. (1991). Analysis of phase distribution and turbulence in dispersed particle/liquid flows, *Chem. Eng. Comm.*, *174*, 85-113.

[22] Anglart, H., & Nylund, O. (1996). CFD application to prediction of void distribution in two-phase bubbly flows in rod bundles, *Nuc. Sci. Eng.*, *163*, 81-98.

[23] Lahey, R. T. Jr., & Drew, D. A. (2001). The analysis of two-phase flow and heat transfer using multidimensional, four field, two-fluid model, *Nuc. Eng. Des.*, *204*, 29-44.

[24] Takagi, S., & Matsumoto, Y. (1998). Numerical study on the forces acting on a bubble and particle, Proceedings of the 3rd International Conference on Mulitphase Flow, Lyon, France.

[25] Tomiyama, A. (1998). Struggle with computational bubble dynamics, Proceedings of the 3rd International Conference on Mulitphase Flow, Lyon, France.

[26] Wang, S. K., Lee, S. J., Lahey, R. T. Jr., & Jones, O. C. (1987). 3-D turbulence structure and phase distribution measurements in bubbly two-phase flows, *Int. J. Multiphase Flow*, *13*, 327-343.

[27] Kurul, N., & Podowski, M. Z. (1990). Multi-dimensional effects in forced convection sub-cooled boiling. Proceedings of the 9th Heat Transfer Conference, Jerusalem, Israel, Hemisphere Publishing Corporation, *2*, 21-26.

[28] Menter, F. R. (1994). Two-equation eddy viscosity turbulence models for engineering applications, *AIAA J.*, *32*, 1598-1605.

[29] Cheung, S. C. P., Yeoh, G. H., & Tu, J. Y. (2007). On the modeling of population balance in isothermal vertical bubbly flows – Average bubble number density approach, *Chem. Eng. Processing*, *46*, 742-756.

[30] Frank, T., Shi, J., & Burns, F. A. D. (2004). Validation of Eulerian multiphase flow models for nuclear safety application, Proceedings of the 3rd Symposium on Two-Phase Modeling and Experimentation, Pisa, Italy.

[31] Sato, Y., Sadatomi, M., & Sekoguchi, K. (1981). Momentum and heat transfer in two-phase bubbly flow-I. *Int. J. Multiphase Flow*, *7*, 167-178.

[32] Patankar, S. V. & Spalding D. B. (1972). A calculation procedure for heat, mass and momentum transfer in three-dimensional parabolic flows. *Int. J. Heat Mass Transfer*, 15, 1787-1806.

[33] Karema, H. & Lo, S. (1999). Efficiency of interphase coupling algorithms in fluidized bed conditions. *Comp. & Fluids*, 28, 323-360.

[34] Karema, H. (2002). Numerical treatment of interphase coupling and phasic pressures in multi-fluid modeling. PhD Thesis, VTT, Technical Research Centre of Finland, Espoo.

[35] Del Valle, V. H. M., & Kenning, D. B. R. (1985). Subcooled flow boiling at high heat flux, *Int. J. Heat Mass Transfer*, *28*, 1907-1920.

[36] Graham, R. W., & Hendricks, R. C. (1967). Assessment of convection, conduction and evaporation in nucleate boiling, NASA TND-3943.

[37] Judd, R. L., & Hwang, K. S. (1976). A comprehensive model for nucleate pool boiling heat transfer including microlayer evaporation, ASME J. Heat Transfer, 98, 623-629.

[38] Steiner, H., Kobor, A., & Gebhard, L. (2005). A wall heat transfer model for subcooled boiling flow, *Int. J. Heat Mass Transfer, 48*, 4161-4173.

[39] Basu, N., Warrier, G. R., & Dhir, V. K. (2005). Wall heat flux partitioning during subcooled flow boiling: Part I – Model development, *ASME J. Heat Transfer, 127*, 131-140.

[40] Basu, N., Warrier, G. R., & Dhir, V. K. (2005). Wall heat flux partitioning during subcooled flow boiling: Part II – Model validation, *ASME J. Heat Transfer, 127*, 141-148.

[41] Warrier, G. R., & Dhir, V. K. (2006). Heat transfer and wall heat flux partitioning during subcooled flow nucleate boiling, *ASME J. Heat Transfer, 128*, 1243-1256.

[42] Bowring, R. W. (1962). Physical model based on bubble detachment and calculation of steam voidage in the subcooled region of a heated channel. *Report HPR-10*, Institute for Atomenergi, Halden, Norway.

[43] Kenning, D. B. R., & Del Valle, V. H. M (1981). Full developed nucleate boiling: overlap of areas of influence and interference between bubble sites, *Int. J. Heat Mass Transfer, 24*, 1025-1032.

[44] Tu, J. Y., & Yeoh, G. H. (2002). On numerical modelling of low-pressure subcooled boiling flows. *Int. J. Heat Mass Transfer*, 45, 1197-1209.

[45] Yeoh, G. H., & Tu, J. Y. (2004). Population balance modeling for bubbly flows with heat and mass transfer, *Chem. Eng. Sci., 59*, 3125-3139.

[46] Yeoh, G. H., & Tu, J. Y. (2005). Thermal hydraulic modeling of bubbly flows with heat and mass transfer, *AIChE J., 51*, 8-27.

[47] Tolubinsky, V. I. and Kostanchuk, D. M. (1970). Vapor bubbles growth rate and heat transfer intensity at subcooled water boiling, Fourth International Heat Transfer Conference, 5, Paper No. B-2.8, Paris, France.

[48] Unal, H. C. (1976). Maximum bubble diameter, maximum bubble growth time and bubble growth rate, *Int. J. Heat Mass Transfer, 19*, 643-649.

[49] Fritz, W. (1935). Berechung des Maximalvolumes von Dampfblasen, *Phys. Z., 36*, 379.

[50] Kocamustafaogullari, G., & Ishii, M. (1995). Foundation of the interfacial area transport equation and its closure relations, *Int. J. Heat Mass Transfer, 38*, 481-493.

[51] Cole, R. (1960). A photographic study of pool boiling in the region of the critical heat flux, *AIChE J., 6*, 533-542.

[52] Ivey, H. J. (1967). Relationships between bubble frequency, departure diameter and rise velocity in nucleate boiling, *Int. J. Heat Mass Transfer, 10*, 1023-1040.

[53] Stephan, K. (1992). Heat transfer in condensation and boiling, Springer-Verlag, NewYork.
[54] Peebels, F. N. and Garber, H. J. (1953). Studies on the motion of gas bubbles in liquids, *Chem. Eng. Prog.*, *49*, 88-97.
[55] Jakob, M. (1949). Heat Transfer, 1, Chapter 29, Wiley and Sons, New York.
[56] Lemmert, M, & Chwala, J. M. (1977). Influence of flow velocity on surface boiling heat transfer coefficient. Academic Press and Hemisphere, New York and Washington.
[57] Končar, B., Kljenak, I., & Mavko, R. (2004). Modeling of local two-phase flow parameters in upwards subcooled flow boiling at low pressure, *Int. J. Heat Mass Transfer*, *47*, 1499-1513.
[58] Basu, N., Warrier, G. R., & Dhir, V. K. (2002). Onset of nucleate boiling and active nucleation site density during subcooled flow boiling, ASME J. Heat Transfer, 124, 717-728.
[59] Hibiki, T., & Ishii, M. (2003). Active nucleation site density in boiling systems, *Int. J. Heat Mass Transfer*, *46*, 2587-2601.
[60] Nylund, O., Becker, K. M., Eklund, R., Gelius, O., Haga, I., Hansson, P. T., Hernbord, G., & Åkerhielm, F. (1967). Measurements of hydrodynamics characteristics, instability thresholds, and burnout limits for 6-rod clusters in natural and forced circulation, FRIGG_1 Report, ASEA and AB Atomenergi, Sweden.
[61] Zeitoun, O., & Shoukri, M. (1996). Bubble behavior and mean diameter in subcooled flow boiling, *ASME J. Heat Transfer*, *118*, 110-116.
[62] Zeitoun, O., & Shoukri, M. (1996). Axial void fraction profile in low pressure subcooled flow boiling, *Int. J. Heat Mass Transfer*, *40*, 867-879.
[63] Donveski, B., & Shoukri, M. (1989). Experimental study of subcooled flow boiling and condensation in an annular channel, Thermofluids Report No. ME/89/TF/R1, Department of Mechanical Engineering, McMaster University, Hamilton, Ontario, Canada.
[64] Dimmick, G. R., & Selander, W. N. (1990). A dynamic model for predicting subcooled void: experimental results and model development, EUROTHERM Seminar #16, Pisa, Italy.
[65] Lee, T. H., Park. G.-C, & Lee, D. J. (2002). Local flow characteristics of subcooled boiling flow of water in a vertical annulus, *Int. J. Multiphase Flow*, *28*, 1351-1368.
[66] Yun, B. J., Park, G.-C., Song, C. H., & Chung, M. K. (1997). Measurements of local two-phase flow parameters in a boiling flow channel, Proceedings of the OECD/CSNI Specialist Meeting on Advanced Instrumentation and Measurement Techniques.

[67] Tu, J. Y., Yeoh, G. H., Park, G.-C., & Kim, M.-O. (2005). On population balance approach for subcooled boiling flow prediction, *ASME J. Heat Transfer, 127,* 253-264.

[68] Krepper, E., Končar, B., & Egorov, Y. (2007). CFD modelling of subcooled boiling – concept, validation and application to fuel assembly design, *Nuc. Eng. Des., 237,* 716-731.

[69] Bonjour, J., & Lallemand, M. (2001). Two-phase flow structure near a heated vertical wall during nucleate pool boiling, *Int. J. Multiphase Flow, 27,* 1789-1802.

[70] Prodanovic, V., Fraser, D., & Salcudean, M. (2002). Bubble behaviour in subcooled flow boiling of water at low pressures and low flow rates, *Int. J. Multiphase Flow, 28,* 1-19.

[71] Gopinath, R., Basu, N. & Dhir, V. K. (2002) Interfacial heat transfer during subcooled flow boiling, *Int. J. Heat Mass Transfer, 45,* 3947-3959.

[72] Luo, H., & Svendsen, H. (1996). Theoretical model for drop and bubble break-up in turbulent dispersions, *AIChE J., 42,* 1225-1233.

[73] Prince, M. J, & Blanch, H. W. (1990). Bubble coalescence and break-up in air-sparged bubble column, *AIChE J., 36,* 1485-1499.

[74] Chesters, A. K., & Hoffman, G. (1982). Bubble coalescence in pure liquids, *Appl. Sci. Res., 38,* 353-361.

[75] Rotta, J. C. (1972). Turbulente Stromungen. Stuttgart: B. G. Teubner, 1972.

[76] Fleischer, C., Becker, S., & Eigenberger, G. (1996). Detailed modeling of the chemisorption of CO_2 into NaOH in a bubble column, *Chem. Eng. Sci., 51,* 1715-1724.

[77] Sateesh, G., Sarit, K. D., & Balakrishnan, A. R. (2005). Analysis of pool boiling heat transfer: effect of bubbles sliding on the heating surface, *Int. J. Heat Mass Transfer, 48,* 1543-1553.

[78] Yeoh, G. H., & Tu, J. Y. (2005). A unified model considering force balances for departing vapour bubbles and population balance in subcooled boiling, *Nuc. Des. Eng., 235,* 1251-1265.

[79] Klausner, J. F., Mei, R., Bernhard, D. M., & Zeng, L. Z. (1993). Vapor bubble departure in forced convection boiling, *Int. J. Heat Mass Transfer, 36,* 651-662.

[80] Zeng, L. Z., Klausner, J. F., Bernhard, J. F., & Mei, R. (1993). A unified model for the prediction of bubble detachment diameters in boiling systems – II. Flow boiling, *Int. J. Heat Mass Transfer, 36,* 2271-2279.

[81] Zuber, N. (1961). The dynamics of vapor bubbles in nonuniform temperature fields, *Int. J. Heat Mass Transfer, 2,* 83-98.

[82] Maity, S. (2000). Effect of velocity and gravity on bubble dynamics, M.S. Thesis, University of California, Los Angeles, U.S.A.

[83] Hsu, Y. Y., & Graham, R. W. (1976). Transport process in boiling and two-phase systems, Hemisphere Publishing Company, Washington.

[84] Bibeau, E. L., & Salcudean, A. (1994). A study of bubble ebullition in forced convective subcooled nucleate boiling at low pressure, *Int. J. Heat Mass Transfer*, *37*, 2245-2259.

[85] Winterton, R. H. S. (1984). Flow boiling: prediction of bubble departure, *Int. J. Heat Mass Transfer*, *27*, 1422-1424.

[86] Clift, R., Grace, J. R., & Weber, M. E. (1978). Bubbles, drops and particles, Academic Press, New York.

INDEX

A

AC, viii, 45
accounting, 4, 52
accuracy, 15
air, 8, 41, 78
algorithm, 16, 21
alternative, 24
application, 16, 25, 35, 49, 60, 71, 74, 75, 78
attention, 25
availability, 55
averaging, 3

B

behavior, 8, 31, 77
birth, 42
boiling, iv, vii, viii, xii, 1, 2, 3, 4, 6, 7, 8, 11, 12, 17, 19, 20, 21, 23, 24, 25, 29, 31, 35, 36, 37, 39, 43, 44, 45, 46, 49, 50, 51, 55, 56, 58, 61, 67, 71, 73, 74, 75, 76, 77, 78, 79
breakage rate, 42
bubble, vii, viii, ix, x, xi, xii, 1, 2, 3, 4, 8, 9, 20, 21, 23, 25, 26, 29, 31, 34, 35, 37, 38, 39, 43, 44, 45, 46, 47, 48, 49, 50, 51, 52, 53, 54, 55, 56, 57, 58, 59, 61, 62, 64, 67, 68, 71, 74, 75, 76, 78, 79
bubbles, vii, viii, x, 1, 3, 4, 6, 7, 8, 19, 20, 23, 25, 35, 36, 37, 39, 40, 41, 44, 45, 46, 49, 50, 51, 60, 61, 64, 67, 71, 76, 77, 78
Bubbles, 40, 79
burnout, 77

C

California, 79
Canada, 73, 74, 77
capacity, 56
cavities, 1, 25, 44, 57
CD, viii, 53
cell, 17, 34, 35
CFD, vii, 1, 3, 19, 23, 35, 43, 74, 78
chemisorption, 78
Chicago, 73
circulation, 77
CL, viii, 53
classes, ix, x, 42, 43, 44, 45, 58, 59, 64
classified, vii, 1
closure, 4, 25, 26, 29, 31, 35, 76
clusters, 77
Co, 16, 17
CO_2, 78
collisions, 40, 46
commercial, 19, 23

Index

complexity, 57, 71
components, 3, 19, 41
computation, 16
computational fluid dynamics, vii, 1, 12, 16
computers, 12
concentrates, 52
concentration, vii, viii, 25, 59, 62, 63, 67, 69, 71, 73
condensation, xi, 3, 4, 6, 31, 34, 37, 38, 39, 43, 44, 45, 46, 47, 48, 62, 67, 74, 77
conduction, x, 3, 19, 49, 50, 51, 56, 75
conductivity, xi
Congress, iv
conservation, 3, 5, 6, 11, 13, 16, 58
construction, 12, 16
contact time, xi, 41
continuity, 7, 12, 44, 45
control, viii, x, xi, xii, 13, 14, 15, 45, 59
controlled, 55
convection, x, 2, 20, 22, 51, 73, 75, 78
convective, vii, 15, 24, 49, 50, 73, 79
convergence, 17
conversion, 16
cooling, 20
correlation, 6, 8, 21, 22, 23, 24, 25, 26, 35, 56, 57, 73
correlations, 2, 3, 4, 21, 23, 25, 37, 49, 58, 71
coupling, 16, 75
covering, 2, 12, 29
Cp, viii
cross-sectional, viii, 34, 45
CTD, viii

D

database, 60
death rate, viii, 42, 44, 46, 48
definition, 1, 15, 20
degree, ix, 4, 57
density, ix, x, xi, 3, 24, 25, 26, 29, 31, 42, 43, 44, 63, 75, 77
detachment, 21, 76, 78
differentiation, 55
diffusion, xii, 11, 12, 13, 15, 55
diffusivity, xi, 55

dispersion, viii, ix, 4, 7, 8, 40
distribution, vii, ix, xii, 1, 4, 7, 8, 25, 26, 31, 34, 35, 39, 41, 42, 43, 44, 45, 55, 62, 67, 71, 74, 75
divergence, 12
division, 12
dominance, 15

E

electronic, iv
electrostatic, iv
energy, ix, 3, 4, 5, 7, 8, 12, 16, 19, 25, 58
energy transfer, 3
engineering, 75
Enhancement, 50
equilibrium, 1, 4
estimating, 55
Eulerian, 3, 64, 74, 75
evaporation, x, 3, 7, 19, 20, 50, 75
evolution, 34, 35, 67
experimental condition, 58, 63, 71
expert, iv

F

failure, 31
film, ix, 40, 41
film thickness, ix, 41
finite volume, 12, 13, 58
finite volume method, 12, 13
Finland, 75
flexibility, 12
flooding, ix, 57
flow, vii, x, 1, 2, 3, 4, 6, 7, 8, 11, 12, 16, 17, 19, 21, 23, 24, 25, 29, 31, 34, 35, 36, 38, 39, 43, 44, 45, 49, 51, 52, 53, 58, 59, 61, 71, 73, 74, 75, 76, 77, 78
flow field, 34
flow rate, 78
fluid, vii, x, 1, 3, 4, 5, 7, 8, 11, 12, 16, 24, 25, 39, 51, 55, 63, 71, 74, 75
fluidized bed, 75
focusing, 29

relationship, 20, 21, 23, 24, 25, 26, 29, 31, 34, 49, 64
relationships, vii, 2, 21, 24, 25, 29, 31, 35, 37, 49, 53, 60
resolution, 12, 42
resources, 42, 64
Reynolds, x, 8, 29, 53
Reynolds number, x, 8, 29, 53
rods, 26, 35

S

safety, 75
saturation, xii, 1, 61
scalar, ix, 42, 44
selecting, 11
separation, 64
series, 42
services, iv
shear, viii, ix, xii, 7, 8, 9, 40, 46, 52, 53, 55, 60
simulation, 35, 74
simulations, 35, 45, 58, 74
sites, x, xii, 1, 44, 49, 51, 61, 76
SME, 75
solutions, 4, 16
spatial, x, 42, 43
specific heat, viii
speed, 37, 46, 64
stages, viii, 16, 40, 67, 71
stress, 8
superposition, 19
suppression, 24
surface area, 13, 40, 41, 45, 49
surface tension, ix, xi, 7, 21, 23, 52, 60
Sweden, 77
symmetry, 29, 59
systems, 3, 4, 17, 40, 41, 73, 77, 78, 79

T

temperature, viii, x, xi, 1, 20, 21, 38, 51, 56, 61, 78
tension, ix, xi, 7, 21, 23, 52, 60

TF, 77
theoretical, 58
theory, vii, 50, 71
thermal, xi, 3, 25, 50, 55, 56
thermodynamic, 1
three-dimensional, 12, 59, 75
thresholds, 77
time, x, xi, 12, 13, 14, 23, 37, 41, 42, 43, 45, 51, 55, 56, 64, 76
tracking, 37
transfer, vii, ix, x, 1, 2, 3, 4, 6, 7, 25, 35, 49, 50, 73, 74, 75, 76, 77, 78
transformation, 71
transition, 2
transparent, 31
transport, 8, 11, 12, 13, 14, 16, 17, 42, 43, 58, 76
transport processes, 11
travel, 64
trend, 59, 62
turbulence, xii, 8, 9, 25, 40, 41, 74, 75
turbulent, viii, ix, xi, 4, 7, 8, 9, 12, 20, 25, 40, 41, 53, 59, 74, 78

U

uncertainty, 49
uniform, 4, 39, 45

V

validation, 76, 78
values, 11, 20, 24, 60, 64, 67
vapor, vii, viii, ix, xi, xii, 1, 3, 5, 6, 7, 8, 9, 19, 37, 44, 45, 52, 54, 55, 57, 59, 60, 63, 64, 65, 71, 78
variable, x, xi, xii, 11, 40
variables, 11, 42
vector, ix, x, xi, 42
velocity, vii, x, xi, 7, 16, 20, 22, 23, 41, 42, 45, 53, 55, 56, 63, 64, 65, 66, 71, 74, 76, 77, 79
viscosity, xi, 9, 25, 75
visual, 31

W

waiting times, 60, 67
wall temperature, 20
Washington, 73, 77, 79
water, 8, 26, 41, 55, 73, 76, 77, 78
weakness, vii, 31, 71
wettability, 57

Y

yield, 60

Z

Zone 1, 27
Zone 2, 27